P. Müller   H. von Storch

Computer Modelling in Atmospheric and Oceanic Sciences

Springer

Berlin
Heidelberg
New York
Hong Kong
London
Milan
Paris
Tokyo

Peter Müller
Hans von Storch

Computer Modelling in Atmospheric and Oceanic Sciences

Building Knowledge

Foreword by Klaus Hasselmann

with 89 Figures, some in color, and 3 Tables

 Springer

Professor Dr. Peter Müller
Department of Oceanography
University of Hawaii
Honolulu
Hawaii 96822
USA
Email: pmuller@hawaii.edu

PROFESSOR DR. HANS VON STORCH
Institut für Küstenforschung
GKSS Forschungszentrum
Postfach
21502 Geesthacht
Germany
Email: storch@gkss.de

ISBN 3-540-40478-3 Springer-Verlag Berlin Heidelberg New York

Library of Congress Control Number: 2004106663

Springer-Verlag Berlin Heidelberg New York
Springer-Verlag is a part of Springer Science+Business Media GmbH

springeronline.com

© Springer-Verlag Berlin Heidelberg 2004
Printed in Germany

Camera ready by authors
Cover design: E. Kirchner, Heidelberg
Printed on acid-free paper   32/3141/as   5 4 3 2 1 0

Foreword

Computer modeling pervades today all fields of science. For the study of complex systems, such as the environment, it has become an indispensable tool. But it is also a tool that is often misunderstood and misinterpreted. These dangers are particularly pronounced in the environmental sciences, an area of interest and concern not only to scientists, but also to the general public, the media, policy makers and powerful interest groups. We cannot experiment with our planet. The only quantitative tool available for the assessment of the impact of our actions today on the future environment and living conditions of later generations is numerical modeling. The better the general understanding of the potential and limitations of numerical models, the better the chances for a rational analysis and discussion of environmental problems and policies. But in addition to the more recent political issue of human impacts on the environment, numerical models play an important role for the forecasting of natural environmental variability, such as tides and storm surges or the weather, or for the interpretation of environmental changes in the past, such as the relation between the Late Maunder Minimum of the sunspot cycle from 1675 to 1710 and the winter half year cooling at the end of the 17th century.

The reasons for misunderstandings and misinterpretations of numerical model results are manifold. The more complicated a model, the more difficult is the explanation of the model structure to the non-expert, and the more readily will the gullible be inclined to over-interpret the model results, while the sceptic may simply shrug his shoulders and comment "trash in, trash out". Conversely, simple models are necessarily based on idealizations and hypotheses that are equally difficult to assess and appreciate for somebody who has not been involved in the detailed analysis of the many different, generally complementary, ways of viewing and reducing a complex system.

Apart from the basic difficulty of conveying the structure of a numerical model to the non-expert (what is established science, what are the hypotheses, what is the data base, what are the numerical approximations?), many misunderstandings arise from the many diverse applications of models. In the environmental sciences, numerical models are used to analyze, understand, infer, initialize, assimilate, project or predict, to name only a few categories, and the assessment of model performance depends strongly on the relevant application type. The performance of a numerical weather model, for example, is judged on its ability to predict detailed weather patterns 3 to 15 days in advance. Whether the model is able to compute the global climate, given the greenhouse gas emissions, 100 years from now, is irrelevant. Conversely, a climate model used for such projections need not necessarily yield reliable short-term weather predictions. Furthermore, the reliability of long term climate projections will generally depend more strongly on the assumed greenhouse gas emissions than on the climate model itself. Nevertheless, the assessment of model performance depends in both cases on the confidence in the basic processes represented in the models, many of which will be common to both classes of model. Thus some degree of cross-coupling in model assessment is inevitable.

But how can confidence be established in the realistic representation of the basic processes of a complex model that, as modelers themselves often complain, are just as complicated and difficult to understand as the real world which they are designed to emulate? Apart from long experience, which is difficult to convey and may raise suspicions of long-grown professional myopia, a standard approach is model reduction: the simplification of a model to a level at which the internal workings of the model do indeed begin to become transparent and understandable. But then one must demonstrate that one has not thrown the baby out with the bath water, that the model world that one now understands is still a reasonable approximation of the real world.

Fortunately, it is not the task of a foreword writer to answer such difficult questions, but only to congratulate the authors on addressing these thorny issues in both depth and width and with a truly commendable clarity. In their presentation, Peter Müller and Hans von Storch draw on a wealth of illuminating examples from their personal experience in developing and working with numerical models in a wide variety of applications, ranging from the transport and deposition of lead in Europe, to tides, predictions of El Niño and state-of-the-art weather forecasts. The scientific development is interwoven with stimulating observations on the philosophy of validation or the public perception of environmental problems. Both the environmental modeler and the non-expert in the field will find the book a rewarding and broadening experience.

Kiebitzreihe, October 2003   Klaus Hasselmann

Preface

Computer models are a major tool in modern environmental systems analysis. These models are based on the integration of a set of approximate dynamical equations. They provide detailed information about the state and evolution of the system. They serve as a substitute or virtual reality that is employed for simulation and experimentation impossible with the real system. These "quasi-realistic" models must be distinguished from "cognitive" models that strive for maximum reduction of complexity.

Numerical experiments with quasi-realistic models allow the derivation or testing of hypotheses. Simulation of actual and hypothetical regimes provide a basis for understanding, analysis and management of a system. Quasi-realistic models share with reality the property that they do not offer immediate dynamical insight. They require further statistical and dynamical analyses to yield understanding. Models are validated by model-data comparison in parameter ranges for which observational data exist. Their usefulness, appeal and danger lie, however, in the option of applying them outside the validated parameter, or configuration, range.

Important issues in quasi-realistic modeling are:

- The reduction of model output to useful scientific and practical knowledge.
- The fidelity and trustworthiness of models when applied beyond the validated parameter range.
- The influence of the experimenter on the virtual model reality.

The main conclusions of the book are:

- Models in general and quasi-realistic models in particular are major tools to expand our knowledge about environmental systems.
- Models of environmental systems are neither true or untrue, nor valid or invalid per se; rather they are adequate or inadequate, useful or not useful to answer specific questions about the system. They describe larger or smaller parts of reality, but never all of reality.

- The validation of models by comparison with data is an important prerequisite for applying models. Validation does not provide new insight about the system, but only insight about the model. New knowledge about the system is gained by applying models to new situations outside the validated range, with all the risks that such an extrapolation entails.
- Models are a tool only. They do not disclose new knowledge by themselves. The act of generating new knowledge still depends on the modeler, on his or her skill in setting up the simulation or experiment, in asking the right questions, and in applying the right methods to analyze the model output.

These points are illustrated by numerous examples drawn from atmospheric and oceanic sciences, especially from efforts to model oceanic tides and the climate system.

The book is organized as follows: In Chap. 1 all the main topics of the book are sketched: the epistemological role of models, the characteristic properties of environmental systems, the two systems, tides and climate, from which we draw most of our examples, the components of quasi-realistic computer models, typical applications of such models, and the main technical, philosophical, economic and social issues. This introduction is non-technical and self-contained. Chapter 2 describes the components of quasi-realistic computer models in detail. Validation, data assimilation and calibration, i.e., the combination of models and observational data are discussed in Chap. 3. The dynamics of tides and climate are further elaborated upon in Chap. 4. The different applications of environmental models, either for applied purposes or for basic research, are discussed in Chaps. 5 and 6. The conclusions are given in Chap. 7. Technical and mathematical details are relegated to a set of appendices which cover dynamics, numerics, statistics and data assimilation.

Choosing an appropriate title for this monograph was not easy. We originally called the book "Computer Modeling in Environmental Sciences" but then opted for "Computer Modeling in Atmospheric and Oceanic Sciences" since all our examples are drawn from these two sciences, in which we have actively been involved. But our general discussions apply to other environmental sciences as well, and we hope to reach readers from these other sciences. Also the subtitle of the book "Building Knowledge" needs an explanation. Alternatives to "building" were "construction" and "acquiring". We wanted to express our understanding that knowledge is nothing absolute, drawn from a firm stock of objective truth. Instead it is the process of constructing skilful explanations for complex phenomena. Of course, this process is not an arbitrary process that results in arbitrary insights. It is a rational process that is guided by continuous comparisons with new observations – but it is *also* guided by our contemporary understanding of the governing dynamics. Thus, understanding is developed by consistently combining previous knowledge and new observations. Eventually, we decided that the term "building" would be best to describe this process, because it describes the fact that any new knowledge *builds* upon previously acquired knowledge. Of course, this process is not foolproof and may lead into dead ends and into inadequate understandings, when the preconceived knowledge is weighted too strongly, and conflicts with new observations are downplayed – a phenomenon well documented in Ludwig Fleck's [38] analysis "Genesis and Development of a Scientific Fact" and Thomas Kuhn's [90] analysis "The Structure of Scientific Revolutions". But the process of building knowledge is at the heart of science and is what makes science such an exciting and rewarding human enterprise.

This book is about one tool to build knowledge: quasi-realistic computer models.

Acknowledgements

Many people contributed to this book in various ways. We would like to thank especially: Stephan Bakan, Lennart Bengtsson, Joachim Biercamp, Grant Branstator, Guy Brasseur, Shailen Desai, Gary Egbert, Josef Egger, Wolfgang Engel, Beate Gardeike, Martin Guiles, Gabriele Hegerl, Diane Henderson, Heinz-Dieter Hollweg, Ben Kirtman, Werner Krauß, Chris Landsea, Mong-Ming Lu, Ernst Maier-Reimer, Jochem Marotzke, Andrei Natarov, Bo Qiu, Jürgen Sündermann, Huug van den Dool, Jin-Song von Storch, Mary Tiles, Hans Volkert, Yuqing Wang, and Francis Zwiers.

Layout: Sönke Rau

Contents

Appendices

1

Introduction

Know thy tools!

"Models" have become an indispensable tool in environmental sciences. The impossibility of conducting experiments with real environmental systems forces scientists to rely on the substitute reality provided by models, especially computer models. For most environmental scientists it seems intuitively clear what models are, but a closer examination reveals many different views. Indeed, curricula in meteorology, oceanography and other environmental sciences usually do not include classes that discuss the role and utility of models in a systematic manner. Instead, young scientists are confronted with a large variety of models, whose limits are not properly defined and whose specific purposes are not explicated. It is the purpose of this book to fill this gap.

The book is based on the premise that the optimal and responsible use of computer models requires that one understands

- what models are,
- what they are based on,
- how they function,
- what their limitations are,

and then, most importantly,

- how models can be used to generate *new* knowledge about the environmental system.

This understanding is important since far-reaching decisions about the environment are based on computer modeling; and a computer model can be the source of valuable information and significant insight in the hands of the able scientist and the source of delusion and misconceptions in the hands of the dilettante.

There exist, admittedly, many books about modeling, but most of them consider modeling either as a technical task, dealing with the manipulation of differential equations and numerics (e.g., [171]), or consider the classical models of physics (e.g., [61], [118]) or economy (e.g., [117]). We, on the other hand, will discuss primarily quasi-realistic computer models that create a substitute reality. These models provide detailed information about the state and evolution of an environmental system. They can be used for simulations and experiments that are impossible with the real system.

In this Introduction we sketch the main aspects of computer modeling in environmental sciences. In Sect. 1.1 we reflect upon the word "model" and the epistemological role of models. In Sect. 1.2 we define environmental systems and discuss some of their relevant properties. The systems *tides* and *climate* from which we draw most of our examples are introduced in Sects. 1.3 and 1.4. The components of quasi-realistic computer models are described in Sect. 1.5, different applications of these models in Sect. 1.6, and fundamental technical, philosophical, economic and social issues in Sect. 1.7. We will come back to all these topics in the remainder of the book.

1.1 Models

The word "model" has many different meanings. The "American Heritage Talking Dictionary" offers:

1. A small object, usually built to scale, that represents in detail another, often larger object.
2. a. A preliminary work or construction that serves as a plan from which a final product is to be made: a clay model ready for casting. b. Such a work or construction used in testing or perfecting a final product: a test model of a solar-powered vehicle.
3. A schematic description of a system, theory, or phenomenon that accounts for its known or inferred properties and may be used for further study of its characteristics: a model of generative grammar; a model of an atom; an economic model.
4. A style or design of an item: My car is last year's model.
5. One serving as an example to be imitated or compared: a model of decorum. See note at ideal.
6. One that serves as the subject for an artist, especially a person employed to pose for a painter, sculptor, or photographer.
7. A person employed to display merchandise, such as clothing or cosmetics.
8. Zoology. An animal whose appearance is copied by a mimic.

To this list we can add many special meanings in various scientific disciplines: For instance, geographers speak of a matrix of numerically given locations and altitudes as a digital terrain model.

In this book we deal, of course, with definition 3 above: "A schematic description of a system, theory, or phenomenon that accounts for its known or inferred properties and may be used for further study of its characteristics".

We define it more specifically: *A model is an analog of a real system in the mind, computer or laboratory.*

We therefore have to distinguish between cognitive, computer and laboratory models. Cognitive models aim at understanding and comprehension and are "simple". Computer and laboratory models are generally quasi-realistic and can be used to create a substitute reality or a virtual laboratory. Emphasis in this book will be on quasi-realistic computer models that create such a substitute reality.

Our definition of the word "model" is similar to its traditional use in the philosophy of science. There, prominent philosophers like Mary Hesse [61] consider models as an "image". These images share some properties with reality. These common properties are called positive analogs. Other model properties are known to not occur in reality or reality has properties that are not represented or misrepresented in the model. These are the negative analogs. Still other properties of the models are neutral analogs. It is not *known* whether or not they are valid in reality. If one hypothesizes that certain neutral analogs are actually positive ones then the model provides *new* hypotheses and, possibly, *new* information about reality.

These philosophers of science also view models, contrary to us, as a preliminary form of a scientific theory. Their models are based on ad hoc formulations that satisfy observed aspects and other knowledge about the process under study. They are often brilliant inspirations of individuals. $E = mc^2$. Such models are "simple". They are cognitive models according to our definition. Often, a model of process A is another process B, which is already understood. The pendulum becomes a model for surface waves. Sound waves become a model for the propagation of light. Different models of the same process eventually lead to a theory of the process. Thus models are a prelude to theory in this school of thought.

Quasi-realistic computer models, on the other hand, are not based on ad hoc choices by brilliant individuals. Instead, they are systematically constructed by a community of scientists. The models are based on theoretical first principles, augmented by phenomenological insights. They require well-informed choices concerning the representation of variables, the approximation of equations, the parameterization of unresolved processes, and the discretization of equations, among many other things. Quasi-realistic environmental models are not precursors to any kind of grand theory about the environment. Instead, they are laboratories.

A term related to model is that of a metaphor, often used in cultural sciences. "The essence of metaphor is understanding and experiencing one kind of thing in terms of another" [91][1]. The relationship between metaphors and models is described by Serres and Farouki [150] as follows:

---

[1] "Our ordinary conceptual system, in terms of which we both think and act, is fundamentally metaphorical in nature. The concepts that govern our thought are not just matters of the intellect. They also govern our everyday functioning, down to the most mundane details. Our concepts structure what we perceive, how

Models offer reproducible partial explanations for phenomena that are too complex to allow for a complete description. Thus, models are not universal; they are valid only in a certain context and if certain assumptions are fulfilled. In this they resemble metaphors which can also only be stretched to a certain point without becoming excessively abstract and, thus, misleading. Models, however, differ from metaphors to the extent that the former are flexible and capable of being adapted to meet scientists' needs, whereas metaphors either fit or do not fit a given situation. In the latter case they have to be discarded.

1.2 Environmental Systems

An environmental system is a segment of the natural environment. Examples are the ocean, the tides, the climate, forests, and the carbon cycle. Such environmental systems are:

- unique,
- complex,
- open, and
- policy relevant.

Here we only discuss the first three properties. The policy relevance is discussed in Sect. 1.4.3 in the context of climate.

Environmental systems are *unique*. No two systems are sufficiently alike that one can infer the properties of one system from the properties of the other system with certainty.

Environmental systems are *complex*. As does the word "model", the word "complex" has many meanings. Here we use the word for systems that have many interacting *degrees of freedom*. For environmental systems these degrees of freedom are also distinct or qualitatively different.

The number of degrees of freedom is the number of values that is needed for a complete description of the system. In the case of the atmosphere the system is described by state variables such as the air velocity, the temperature, the humidity, the air pressure and others (Sect. 4.2.3). These state variables are continuous functions of position. At each position the function can assume different values. The specification of the state variables thus requires an infinite number of values. The system has an infinite number of degrees of freedom. Because the state variables are spatially correlated one may approximate the continuous functions by piecewise constant functions,

---

we get around in the world, and how we relate to other people. Our conceptual system thus plays a central role in defining our everyday's realities. If we are right in suggesting that our conceptual system is largely metaphorical, then the way we think, what we experience, and what we do every day is very much a matter of metaphor." [91]

constant over a finite number of grid cells. Then, the number of degrees of freedom is the number of grid cells times the number of state variables. An alternative is to expand the continuous functions with respect to an infinite set of (orthogonal) basis functions, like sinusoids on the real line or spherical harmonics on the sphere, and truncate the infinite series. Generally, the number of grid cells (or retained basis functions), and hence the number of degrees of freedom, is not well-defined. One usually opts for the smallest number that allows for a sufficiently accurate description of the overall system, but even this smallest number is usually fairly large for environmental systems such as the atmosphere.

The second important aspect of a complex system is that the different degrees of freedom interact or are coupled to each other. The time evolution of one degree of freedom depends not only on itself but also on the other degrees of freedom. "Everything depends on everything else". The coupling is due to the nonlinearities in the dynamics of the system.

A third important aspect of complex environmental systems is that the many interacting degrees of freedom are distinct or qualitatively different. They and their time evolution need to be treated individually. They cannot be related to each other by simple scaling or similarity laws. As a consequence environmental systems exhibit strong *inhomogeneity* and *irregularity*.

As a typical example consider the formation of fronts in coastal seas. It depends on many factors. It depends on radiation which is affected by the distribution of clouds. It depends on the surface mixed layer which is affected by the spatial pattern of winds and gusts. It depends on the rainfall in the catchment which again depends on the presence of clouds. And so on. Thus, many qualitatively different degrees of freedom interact with each other, differently in different locations.

Qualitatively different degrees of freedom are even more pronounced in ecosystems where the interaction between physical, chemical and biological components and processes requires an interdisciplinary approach.

Environmental systems are *open*. They are not isolated. They are under the influence of a variety of factors that are external to them. These external factors can be human influences, astronomical influences, or influences from other components of the environment. The emission of greenhouse gases into the atmosphere by automobiles is a human influence on climate, the Milankovitch cycles are an astronomical influence, and the volcanic emission of sulphate aerosols and radiatively active gases into the atmosphere represents an influence from another environmental component. The bathymetry of the ocean and the coastline, including man's modifications, are also external factors. It is impossible to account for and specify all external factors. The specification of the emission of greenhouse gases, for example, requires the forecasting of human behavior and ingenuity – which can hardly be done. Even the factors that one can account for are usually not completely known. The bathymetry of the ocean, for example, is available only in approximate form.

The openness of the system has a profound consequence. The ability of a model to reproduce observations may be due to the skill of the model or due to the influence of relevant external factors that are neglected in the model. Thus, environmental models cannot be verified[2].

Most of the examples in this book will be drawn from two environmental systems: tides and climate[3]. The tides are a relatively simple system. They are well described by contemporary models. In contrast, climate is a much more complex system. It has more state variables, more degrees of freedom, and stronger nonlinearities. Nevertheless, contemporary quasi-realistic climate models are considered to describe the climate system sufficiently well to allow for scenario building and other applications. In the next two sections we give a general overview of the phenomena and the modeling efforts. In Chap. 4 the dynamics of tides and climate will be discussed in more detail.

1.3 Tides

Tides are waves with very long wavelengths that cause the regular fall and rise of sea level observed at most coastlines.[4] They are caused by the gravitational attraction of the moon and sun. Tidal rhythms have always affected man living at open ocean coasts and controlled their lives (see Fig. 1.1). Vital issues such as navigation and coastal protection depend on knowledge about ocean tides. (The Greeks and Romans knew almost nothing about tides since the Mediterranean Sea is too small for tidal generation. Alexander the Great lost his fleet in India due to unexpected strong ebb currents). Here we discuss briefly the role of the tides in environmental systems and different kinds of modeling approaches.

1.3.1 The Role of Tides Within Environmental Systems

Tides exert significant control on marine systems by

---

[2] With *verification* we mean the assertion that a model provides the right response to forcing for the right reasons. For a detailed discussion refer to [127] and Sect. 3.1.

[3] It would have been a challenge to include as a third example marine ecosystems. Such ecosystems are even more complex and are influenced by even more factors than climate, with nonlinear interactions among all scales and state variables (species, substances). There is not even consensus about the relevant state variables. Marine ecosystem models have not yet matured to a state where they can reliably be used for management decisions. For these reasons we decided to not include this exciting topic and limit ourselves to tides and climate.

[4] The text of this subsection was to a large extent provided by Jürgen Sündermann, Hamburg.

Fig. 1.1. Low and high tide in the Fundy Bay. From Defant [25]

- advecting matter and energy,
- generating turbulence and mixing, and
- providing an amphibic environment with moving coast lines.

In open seas, freely connected to the world ocean, tidal currents are always present, at any depth down to the bottom of the ocean. Tidal currents are periodic, but due to nonlinear effects, which are especially strong in shallow waters, they do not cancel out when averaged over one period. A steady *residual current* is the result (see Fig. 1.2). These residual currents condition

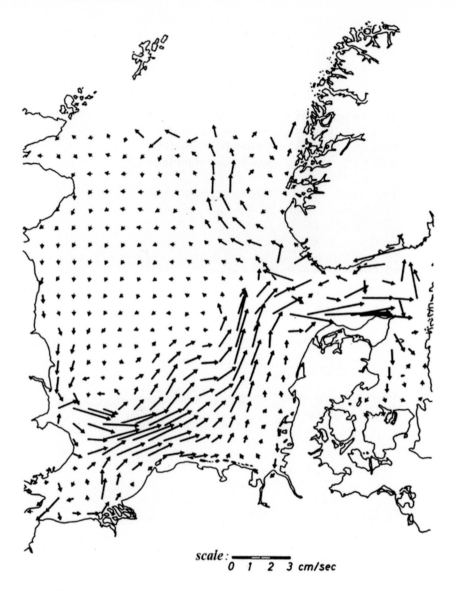

Fig. 1.2. Residual currents of the M_2 tide in the North Sea [12]

the hydrographical, morphodynamical and biological environment and have important implications for the marine environment.

Mixing is the second important process that tides exert on the environment. In coastal seas, the periodic tidal excursions represent mesoscale motions and move water parcels up to 20 km within half a day. When combined with other irreversible processes these excursions can lead to significant horizontal mixing. Second, tidal flows generate a turbulent bottom boundary layer, which is nearly fully mixed. It can extend into a significant part of the water column and destroy thermohaline fronts and prevent the formation of stratification. Indeed, because of strong tides no stable stratification can be established in the southern North Sea even in summer; this has a severe impact on the primary production in this region. In the deep ocean, tidal currents interacting with topography generate *internal* or *baroclinic* tides[5]. They are assumed to be an important part of open ocean vertical mixing.

The morphodynamical evolution of the coastal zone and the recruitment strategies of many marine organisms are conditioned by the tides in many areas.

Oceanic tides are also contained as a prominent signal in many geophysical, geodetical and astronomical records. Tides cause short-period and long-term changes of the gravity field and the shape of the earth, of the length of day and the position of the poles. The respective sciences are therefore very interested in a precise knowledge of ocean tides. Vice versa, the tidal signals in these often extremely accurate measurements (orbits of satellites, Lunar Laser Ranging, Very Long Baseline Interferometry) provide independent terrestrial data for the validation of tidal models.

1.3.2 Different Modeling Approaches

Tidal elevations show an often complicated but always regular behavior. The phases are related to the moon's and sun's position, which has been recognized from early on. From tidal records simple and extremely reliable rules could be established for the forecast of tidal elevations, such as (i) high water (low water) occurs every 12 hours and 25 minutes, (ii) every second high water (low water) is smaller (daily inequality), and (iii) every two weeks tidal ranges are high, in between they are low.

This and other tidal behavior has been investigated by various modeling approaches:

- harmonic analysis,
- hydraulic models,
- equilibrium models, and
- dynamic models.

The harmonic analysis represents the tidal signal as a superposition of harmonic tidal constituents (partial tides) with certain celestial (not oceanic)

---

[5] The tidal currents caused by the gravitational attraction of the moon and sun extend nearly uniformly through the water column. They are therefore often referred to as *external* or *barotropic* tides. In contrast, internal or baroclinic tides are currents of tidal frequency but with significant vertical structure. They are a byproduct of the barotropic tides.

frequencies. The exact amplitudes and phases of the constituents can be obtained by a harmonic analysis of a long-term tidal record. At the beginning of the 20th century machines performed this analysis for a limited number of tidal constituents (Fig. 1.3). Nowadays, harmonic analysis is done using computers.

Fig. 1.3. The mechanical tide simulator operated at the German Hydrographic Institute. From Defant [25]

Tidal elevations can now be forecasted by this harmonic expansion and extrapolation method with sufficient accuracy and reliability for several years in advance. Official tide tables are mostly based on this method. There remains a certain deficiency at shallow water coasts: here the observed tidal curves show a non-sinusoidal shape which cannot be approximated by even a high number of partial tides. Higher harmonics are required.

A hydraulic model (Fig. 1.4) is another, more natural and suggestive, way to model tides. Here the tidal phenomenon is simulated within a hydraulic apparatus, assuming the validity of certain hydrodynamic similarity laws. The currents and elevations in the apparatus are governed, like in nature, by gravity, geometry and friction and are forced, like in nature, at the boundary. For a while such hydraulic models were very popular for engineering applications. But they are restricted to the local scale (usually the Coriolis force is not

Fig. 1.4. The hydraulic tidal model of the Bundesanstalt für Wasserbau in Hamburg. From Sündermann and Vollmers [163]

represented) and require high experimental effort. Today they are mainly used for guiding constructions and for process studies.

Figure 1.4 shows a hydraulic model of Jade Bay, a tidal inlet in the southern German Bight[6]. The bay is approximated to have a circular geometry, with a diameter of 10 km, and a 6.25 km long and 4 km wide channel to the open sea. The bay is assumed to have a uniform undisturbed water level of 15 m. The photograph shows the miniaturized bay, with a reduction factor of 10^{-3} in the horizontal and 10^{-2} in the vertical, seen from the "sea". Thus, the miniaturized model bay has a diameter of 10 m, the channel is 4 m wide and 6.25 m long. The undisturbed water depth is 15 cm. At the entrance to the channel a wave with an amplitude of 1.51 cm and a period of 7.45 minutes is excited (corresponding to 1.51 m and 12 h 25 min's). The water levels and the current speeds and directions are recorded. Figure 1.5 shows the current directions right behind the corner on the left side of the bay shortly before and after high tide, given by a time exposure photograph. In both photographs a cyclonic flow appears, with a net inflow shortly before the high tide, and a net outflow shortly after. Records of current velocities during one tidal period are displayed in Fig. 1.6, together with the result of a numerical model.

---

[6] We discuss this case in some more detail, as we will revisit it in Sect. 4.1.4 and in Sect. 6.1.1.

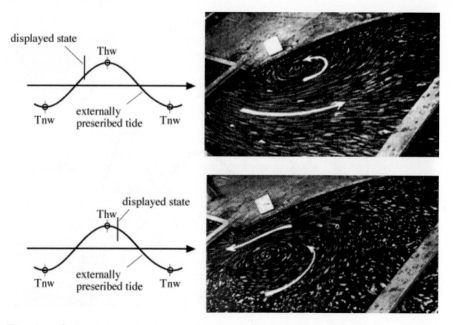

Fig. 1.5. Currents in a simulation with the hydraulic tidal model shown in Fig. 1.4 shortly before and after high tide (see inlet). The area shown is at the corner between the circular bay and the channel, on the right hand side of the photograph in Fig. 1.4. From Sündermann and Vollmers [163]

The most promising approach to model tides is, however, to consider the ocean as a fluid governed by physical laws. These laws are encoded in the basic equations of fluid dynamics (Sect. 2.1). These equations describe quantitatively the response of a fluid to external forcing (astronomical forcing in the case of tides) in a given (geometrical) environment. Assuming that the earth is totally covered by an ocean of uniform depth, Newton was able to calculate the equilibrium response of the ocean to the tide-generating astronomical forces. His model explains the above-mentioned tidal periods and inequalities, and even global tidal ranges. His model is also adequate to understand the spatial structure of the long-period tides (from two weeks upwards; Fig. 1.7a). It fails, however, to explain the complex oscillation structure of the diurnal and semi-diurnal tides (Fig. 1.7b). Newton's "equilibrium model" is too simple. It does not account for the effect that gravity waves propagate with a finite phase speed. The shorter period tidal waves cannot follow the moon's or sun's orbit. Tides become a transient phenomenon. This results in complicated oscillation patterns, which are further modified by the continental boundaries.

Newton's equilibrium model is a cognitive model. It explains in a maximum simplified manner some general features of the tides. However, it cannot be

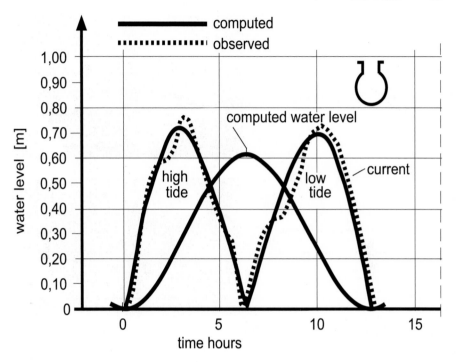

Fig. 1.6. Comparison of tidal currents obtained by the hydraulic model shown in Fig. 1.4 (*dashed*) and the numerical model discussed in Sect. 4.1.4 (*solid*). The simulated water level is also shown. From Sündermann and Vollmers [163]

used for specific planning purposes or for explaining details that arise from realistic topography.

Laplace accounted for the transience of the tides in his "dynamical approach". "Laplace tidal equations" are a set of partial differential equations that form a sound basis for the quantitative description of tides in the world ocean. Their drawback is that they are complicated. They cannot be solved analytically, even for a sphere covered by an ocean of uniform depth. Again, we are faced with the practical limitations of a cognitive model.

Significant progress towards the realistic modeling of the tides was achieved by solving Laplace tidal equations numerically. The differential equations are discretized and turned into a set of algebraic equations. The problem is reduced to the manipulation of a large but finite set of numbers, which can be carried out by computers. One arrives at a numerical model of the tides. In principle, these numerical models can be designed to represent reality ever more accurately. The resolution can be increased. More and more processes can be included, such as bottom friction which is important in the shallow parts of the oceans. The results of such numerical models become closer and

Oscillation system of the Mf

Corange lines (broken) in mm
Cotidal lines (full) in 30⁰

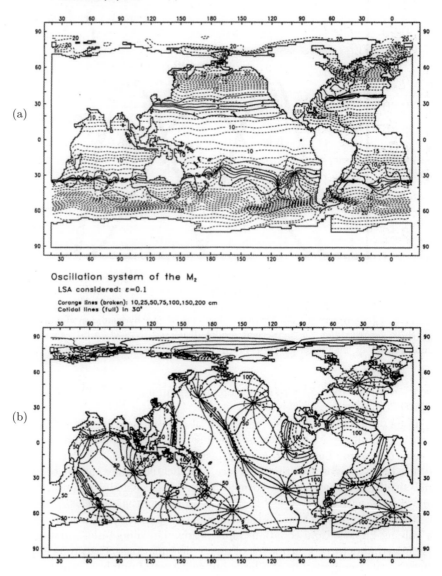

(a)

Oscillation system of the M₂

LSA considered: ε=0.1

Corange lines (broken): 10,25,50,75,100,150,200 cm
Cotidal lines (full) in 30°

(b)

Fig. 1.7. Amplitudes and phases of the fortnightly (**a**) and semi-diurnal M₂ (**b**) ocean tide, simulated with a quasi-realistic numerical model. These simulations have not yet taken into account the processes of tidal loading and self-attraction as in Fig. 5.33; satellite data have also not been assimilated as in Fig. 5.35 [151]

closer to reality. They become *quasi-realistic*[7] models. These models must, however, still be analyzed and interpreted in order to understand the system.

A simulation of Jade Bay with the hydraulic model is compared in Fig. 1.6 with a simulation from a numerical model. The currents are simulated in both models similarly, lending credit to both models. However, the numerical model is more powerful, as it additionally provides information about the water level, which could not be measured in the hydraulic model. We will later see (Sect. 6.1.1) that the numerical model has other advantages over the hydraulic model as well. For this reason, tidal studies are no longer done with hydraulic models.

Quasi-realistic tidal models, on the other hand, are suitable for various practical purposes. They may be used for numerical experiments, e.g., to study the effect of self-attraction of the ocean, for detailed forecasting, and for the analysis of satellite data.

The dynamics of tides is further discussed in Sect. 4.1.

1.4 Climate

The word "climate" is used here to mean the statistics of weather. It thus includes not only mean values but also variances, extreme events and other statistical descriptors. Weather is then a realization of climate.

As tides, climate has in all times been an important constraint for people. In fact, in the not-too-distant past climate was understood solely as the statistics of weather as far as it affects people. There was no "climate" of Antarctica or Mars.

Here we briefly describe the historical development of climate science, discuss qualitatively quasi-realistic climate models and comment on the societal relevance of climate.

1.4.1 Historical Development

The science of climate has developed along two separate lines aimed either at

- describing the geographic variations of climate, or
- understanding climate as a "thermodynamic engine".

These lines have now merged and given way to an interdisciplinary systems analysis approach.

The geographic descriptive approach views climate as a resource or limiting factor. Climate is therefore a local or regional object. In areas where temperatures fall below the freezing point, palm trees do not grow, and the agronomist is advised not to grow citrus trees. Therefore, not surprisingly, climatology as

---

[7] We use the notion *quasi-realistic* instead of *realistic* because models are always approximations and therefore reflect only part of reality.

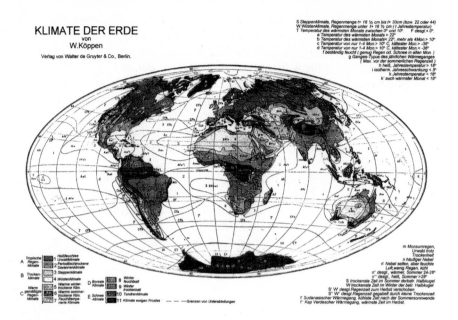

Fig. 1.8. One of Vladimir Köppen's climate maps of the world [88]. The figure caption is in German

a science began with the description of local and regional climates, and the various regional charts were combined into world maps. The global climate was perceived as the sum of regional climates. Famous representatives of this line of research are Vladimir Köppen [88], Eduard Brückner [155] or Julius von Hann [172]. Köppen's maps are even nowadays in use. They classify the surface of the world into climatic zones, which are determined by the amount of precipitation they receive, and the temperature regime. Dynamical quantities such as the wind are considered secondary. One of Köppen's maps is reprinted in Fig. 1.8.

The weather services pursue this traditional line of research by providing "climate normals" for planning purposes in traffic, agriculture, tourism, and for other applications. Such normals are not only mean values derived from, say, a 30-year interval, but also the probability for extreme events (such as 100-year storm surges).

The "thermodynamic engine" approach, on the other hand, aims at understanding why climate is as it is. This approach developed parallel to the geographic descriptive approach over the course of the last centuries. Examples are George Hadley's explanation of the trade wind system[8] (see Fig. 1.9), Immanuel Kant's postulation of a continent south of the Indonesian archipelago

---

[8] For an interesting review of the history of ideas concerning the general circulation of the atmosphere, of which the trade wind system is a part, refer to [103].

Fig. 1.9. Hadley's concept of thermally driven vertical cells, which are deflected by the rotation of the earth, creating the equator-ward directed trade winds [103]

– at that time unknown to Europeans – based on wind observations from merchant vessels, and Arrhenius' [3] hypothesis about the impact of air-borne carbon dioxide on near-surface temperatures. Interestingly, all these theories turned out to be essentially correct, in spite of the severely limited observational evidence that was available to these researchers. As a demonstration, a modern sketch of the general circulation (Fig. 1.10) is juxtaposed to Hadley's 17th century concept. The two figures have many common elements.

The present state of climate is understood to be determined by the condition that the incoming solar or shortwave radiation balances the outgoing thermal or longwave radiation. This principle is encoded in the concept of the *energy balance model*, depicted in Fig. 1.11. In tropical latitudes, the earth's surface and the atmosphere gain more energy from the sun than they lose to space as thermal radiation. The excess energy is transported poleward first by the Hadley Cell and then by horizontal eddies to mid-latitude and polar regions (Fig. 1.10), where less solar radiation is received than lost to space. Thus the atmosphere acts like a thermodynamic engine, heated at the tropical surface, and cooled at mid and high latitudes. This engine is doing work – causing the movement of mass which we call winds. The ocean functions differently. In the present geological period, the deep ocean is cold, close to freezing. Thus, the ocean is stably stratified. Any circulation involving both the upper and lower layers of the oceans requires a mechanism that connects the two layers. Such a mechanism is convection in subpolar regions, where sea ice is formed and the surface waters are cooled by the atmosphere. Both processes cause the ocean water to become denser. Convection sets in. Surface water sinks down to deeper levels and displaces water there, which moves along the *conveyor belt*, as sketched in Fig. 1.12. Thus, the ocean circulation

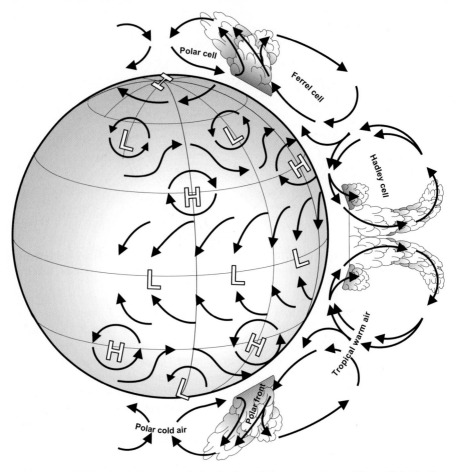

Fig. 1.10. A sketch of the general circulation of the atmosphere. The mid latitudes are characterized by unsteady (but statistically quasi-stationary) states [177]

may also be seen as a thermodynamic engine, with densification taking place at subpolar regions[9]. The work done by this engine takes the form of ocean currents that transport thermal energy from low to high latitudes.

The two approaches to climate, the geographic descriptive and the thermodynamic engine approach, have long been separated. They are now reconciled after meteorology and oceanography began to comprehend their systems, the atmosphere and the ocean, as parts of a giant geophysical machinery, named "climate". This and global observing systems, partly operating from satellites, new biogeochemical techniques for reconstructing past climate states, and,

---

[9] Here we mean the circulation that involves both the upper and lower layers of the ocean. At the ocean surface various circulation patterns also develop in response to wind forcing and bathymetric features.

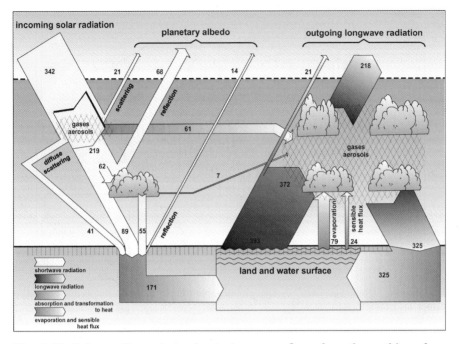

Fig. 1.11. Balance of incoming and outgoing energy fluxes from the earth's surface, the atmosphere and space. Units are W m^{-2} [174]

foremost, the development of quasi-realistic climate models and supercomputers caused a change in the paradigm of climate research, from describing an anthropocentric environment to understanding, modeling and predicting climate, using systems analysis techniques. This development was advanced by the concerns about anthropogenic climate change, which added urgency and need for applying climate-related knowledge.

A major insight of climate research is the understanding that one must distinguish between local, regional, planetary and global climate. Global features are defined as the mean of many local features. The global mean *near-surface temperature*[10] is such a global feature. It can directly be determined from *energy balance models* without any knowledge about local aspects. The major features of the general circulation of the atmosphere, such as the tropical meridional cells and the jet streams associated with baroclinic instability and the formation of storms, constitute the planetary climate. It is fairly well simulated by an aquaplanet, a planet covered entirely by water, without any mountains, coasts and vegetation. The distribution of continents and the presence of the large mountain ranges, such as the Himalaya, the Andes, the Rocky Mountains, Antarctica and Greenland, only modifies this planetary climate.

---

[10] It is defined as the globally averaged air temperature at 2 m height above the ground.

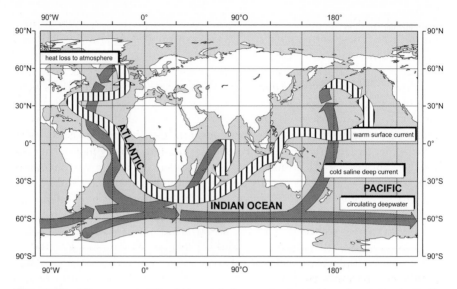

Fig. 1.12. "Conveyor belt" of the global oceanic circulation, involving both the upper and deep ocean

Current "low-resolution" general circulation models describe this planetary climate well. Once the planetary scale state of the atmosphere is set, the regional and local climates emerge as the result of an interplay between the planetary climate state and local physiographic features such as geographical location, local topography, proximity to the ocean and land use [174].

1.4.2 Quasi-realistic Climate Models

When we refer to climate modeling in this book, we mean "quasi-realistic global climate models", which are under construction since the 1960s [107]. Such models attempt to account for as many as possible mostly physical processes, in order to simulate the detailed evolution of the atmosphere, the ocean, the cryosphere and perhaps some other climate components such as biogeochemical cycles or vegetation. These models resolve spatial scales of a few hundred kilometers. Their temporal evolution ranges from days to several hundred and even thousands of years.

We do not consider conceptional models that, for example, try to explain the latitudinal extension of the Hadley Cell, nor idealized models such as energy balance models nor integrated assessment models, which combine dynamically strongly reduced climate models with rudimentary economic models. We also do not consider process models that, for example, study the diurnal circulation in an Alpine valley, the injection of volcanic aerosol into the stratosphere or the formation of fronts in the North Sea. We focus on quasi-realistic models. "Quasi-realistic" climate models are assembled from sub-models of the atmosphere, the ocean, the cryosphere and other climate components. These sub-models are based on dynamical equations. These are usually balance equations for mass, momentum and energy for the respective medium and equations characterizing the medium. After discretization these equations can be solved on computers. Unfortunately, these equations must include the effect of small-scale processes that are not resolved by the model but affect the resolved scales and processes. Such subgridscale processes include convection, cloud formation, precipitation, the absorption and emission of radiation, surface friction, ice formation, run-off and many more. They are labelled "physics" in the jargon of climate modelers. In the energy balance of the atmosphere they often constitute the dominant sources and sinks. Since these subgridscale processes are not explicitly resolved, their effect on the resolved scales must be parameterized. The functional form of "parameterizations" is mostly physically motivated, but the choice of crucial parameters depends on empirical findings and on the "success" of the parameterization scheme when implemented into a climate model. All environmental models contain such parameterizations, often in very disguised form. Not including a process is also a form of parameterization. From an epistemological point of view this is somewhat dissatisfying. However, because all scales of the climate system interact, parameterizations are an indispensable element of climate models.

These quasi-realistic climate models are the only tool for investigating the dynamics of climate and for deriving detailed scenarios. Thus it is of utmost importance for the users of information provided by climate models to be aware of the limits of such models. The two most important limits are:

- Global climate models simulate well-resolved phenomena satisfactorily, but fail for processes and patterns that are hardly or not resolved. "Well resolved" are planetary scales of several hundred and more kilometers, whereas "hardly or not resolved" are regions such as the islands of Hawaii, Israel or the North Sea. Climate models thus do not give any immediate information about detailed ecological and economical impacts of climate change since these require information on regional and local scales.
- Climate models do not forecast. They simulate plausible realizations of a random process. Different initial conditions lead to different trajectories but identical statistics.

It should also be kept in mind that climate research is more than meteorology or oceanography; it is an interdisciplinary systems analysis approach. The underlying dynamics of the climate system is discussed in more detail in Sect. 4.2.

1.4.3 Societal Relevance of Climate

Climate has at all times been perceived by people as an important factor for human life and its organization. Greek philosophers as well as thinkers from the Enlightenment took such influences for granted. Even today, this view seems to be firmly implanted, at least in the western view of the world. More sophisticated, allegedly scientific versions were put forward in the last century by "climate determinists" such as Ellsworth Huntington [66] (see also [156]). These climate determinists went so far as to assert that the appearance of people, their physical and intellectual capabilities and attitudes are determined by the climatic conditions. Similarly, Markham [109] claimed in 1947 that the progress of civilization was steered by peoples' growing ability to create favorable in-door climates, helping England to rise to a global power in spite of its suboptimal climatic conditions. These concepts were formulated so broadly that they could not be falsified [157]. The concepts were actively rejected by social scientists such as Emile Durkheim since the beginning of the 20th century by pushing the axiom that "social issues are explained by social factors". Durkheim's axiom is still strictly accepted by many social scientists today, and this is one of the obstacles to interdisciplinary co-operation between natural and social sciences. A more adequate position, avoiding both extremes of rejecting any role of the natural environment and of assigning it a dominant role, was formulated by the geographer Wilhelm Lauer [96]:

> Climate shapes the theater in which human existence – the history of the human race – takes place, sets borders for that which can happen on the earth, but certainly does not determine what happens or will happen. Climate introduces problems that man has to solve. Whether he solves them, or how he solves them, is left to his fantasy, his will, and his formative activities. Or, expressed in a metaphor: climate does not compose the text for the development drama of mankind, it does not write the movie script, that does man alone.

A somewhat milder form of climatic determinism, for instance represented by Eduard Brückner at the turn of the 20th century, was the hypothesis that changing climatic conditions would increase or decrease the nations' possibilities in dealing with peoples' health, agricultural production or long-range transport [155]. This claim is interesting insofar as the analysis of the vulnerability of societies to climate changes was certainly correct at that time, but proved to be irrelevant within very few decades, simply because of the pharmaceutical progress made, the improved agricultural practices and the large-scale introduction of railroad traffic. Two recent cases, one reported by Reiter [139] about the spread of mosquito-borne diseases (Malaria) and one by Klinenberg [86] about the deadly heat wave in Chicago in 1995, support this conclusion. In both cases it had been hypothesized that climate had a significant impact on health and mortality. A closer analysis, however, revealed that changes in the efficiency of the health system is the key for understanding the different spread of malaria, and that recent adverse social practices (isolation of the poor elderly) was the main cause for the high mortality during the heat wave in Chicago.

On the other hand, single extreme weather events, like wind-, hail- or rain storms have severe impacts on humans and the economy. The same is true for climatic anomalies, like the extended drought during the Dustbowl episode in the 1930s. But also in these cases the impacts are often aggravated by insufficient preparedness and unsustainable use of the environment.

An important aspect is, however, that the political discourse is not so much about the real impact of weather and climate on society, but much more about society's *perception* of weather and climate impacts. And this perception is often rather different from the real impacts. Examples are given by Kempton and Craig [81] and Kempton et al. [80]. One such example is that the past weather is falsely remembered as regular and consistent with the seasons. In the past, the seasons were more clearly separated, but nowadays, they have become more disorderly. This perception has been examined by Rebetez [138] in some detail with respect to "White Christmas" in Switzerland. Thus, scientific knowledge is competing in the public arena with other knowledge claims, which are based on historical, social and cultural constructions.

On the other hand, society is influencing climate, through changing land use and the emissions of anthropogenic substances into the atmosphere. Land use changes mainly affect local and regional scales; emissions of aerosols from burning coal and forests have mainly a regional effect, but the emission of greenhouse gases is leading to changes on the global scale. These anthropogenic changes have until now been rather minor but detectable [197] and are expected to increase at an accelerated speed in the next decades. The impacts of this "Global Warming" are not really understood and will emerge in parallel with other significant changes in economic activity, societal attitudes and social and cultural practices. Thus, some climatic impacts may actually turn out to be insignificant in the future, even if the same impacts today would be markedly harmful. And conversely, present societally insignificant changes may have severe effects in the future. In view of this perspective it is advisable to resort to the precautionary principle and to try to reduce anthropogenic climate change as much as possible without significant social and other repercussions. The Kyoto protocol is considered a first step in this direction, even though it now has a mostly symbolic value and will not directly limit the expected climate change in a significant way.

The term "Global Warming" has become a household term, which needs no explanation when used in the news [169]; however its content is mostly severely misunderstood [80]. Again, public knowledge is not aligned with scientific knowledge. Because of this mismatch, climate science is no longer "normal science" [90] driven by curiosity, but "postnormal science" [42], [11], [144], which is characterized by high uncertainty and high risks. In a postnormal framework, the concept "science speaks to power" is no longer valid [10]; instead, different opinions, nurtured by different knowledge claims, are competing with each other and the attention of the public and, eventually, of the policy makers.

1.5 Quasi-realistic Computer Models

Here we discuss some general characteristics of quasi-realistic computer models. The more technical aspects are discussed in Chap. 2.

The need for quasi-realistic modeling

Quasi-realistic models are the only tool to address a variety of problems in environmental sciences. The main reason is that certain experiments[11] cannot be performed in reality. They are either impossible, impractical or unethical. For global systems such as the tides or the climate one can simply not perform experiments. One cannot change the amount of solar radiation, the position of the moon and sun or the shape of the continents. There are global experiments that we do perform, such as the release of carbon dioxide into the atmosphere, but we would like to anticipate the outcome of this experiment rather than to wait for it. For environmental systems, we also do not have the "controlled conditions" required for meaningful experimentation, because environmental systems are open and are influenced by all kinds of unaccountable factors. The uniqueness also limits experimentation. No two systems are sufficiently alike that results from experiments with one system can be applied to other systems. Even if experiments can be done it may be unethical to perform them. This is particularly the case in the biological and medical sciences. Thus quasi-realistic models are needed.

Dynamics

In this book we consider models that are based on dynamical equations. These dynamical equations represent the laws of nature for the system under consideration. Computer models can also be based on kinematic extrapolation or on analogs. These models are not considered in this book. The assertion that models are based on dynamical equations sounds more innocent than it is. The question is whether the system to be described is characterized by a unique set of dynamical equations "representing the laws of nature". In general the answer will be negative. The need to approximate the laws and include parameterizations implies that different formulations, using different approximations and containing different parameterizations, are possible. This is the major problem in the case of tides and climate. In the case of marine ecosystems one encounters the additional problem that the relevant state variables are not even known[12].

---

[11] Following the American Heritage Talking Dictionary an experiment is: "A test under controlled conditions that is made to demonstrate a known truth, examine the validity of a hypothesis, or determine the efficacy of something previously untried".

[12] In the case of tides, the state variables are the vertically averaged currents and the water elevation. In the case of the atmosphere or ocean, the state variables are the fluid velocity, the temperature, the pressure and the humidity (salinity).

In this book we consider only cases for which it is not a problem to formulate the dynamical equations that represent the laws of nature. These laws state the causes for change and evolution. They are cause and effect statements. The premier example is Newton's second law. It states that the acceleration of a particle is given by the force acting on the particle. The force is the cause. The acceleration is the effect. The field of physics that studies the motion of particles under the action of forces is called dynamics. Here we broaden this meaning and call *dynamics* all natural laws that state a cause and effect relationship. Dynamical models thus view a system as a causal network and ascribe changes of the system as being the effect of particular causes acting on or within the system. The cause and effect relations are called dynamical laws. The dynamical laws for a fluid system, such as the ocean or atmosphere, are the subject of fluid dynamics and are described in standard textbooks like [129], [63] and [45]. A brief summary is offered in Appendix A.

A short discussion of the role of empiricism may be in order at this point. Whether dynamical equations are known or not, the future development of any system may be estimated by means of empirical extrapolation. Extrapolation is applied when an observational record shows sufficient regularity. One then extrapolates this regularity to predict the future evolution of the system. Such extrapolation has been used since ancient times in astronomy, with considerable success. The Greek philosopher Thales correctly predicted a total eclipse of the sun in 589 BC. Extrapolation does not give the causes for the evolution of the system. Today it is used to predict tidal currents and elevations by harmonic analysis of the tidal record (cf. Fig. 1.3). The extrapolations in astronomy and tidal analysis are deterministic. Often an observational record shows irregular behavior but with some structure in it. One set of events is followed often but not always by some other set of events. In this case, statistical methods allow extrapolation and prediction. Statistical extrapolation is uncertain. It does not predict a specific evolution but an ensemble of evolutions. From such an ensemble one can construct the typical or averaged evolution and quantify the uncertainty. Statistical extrapolations also do not give the causes of the evolution. Such extrapolations are not unscientific but are not considered in this book.

Computational formulation

In general, the dynamical equations cannot be solved analytically but must be solved numerically. This requires computational algorithms and codes. The issues that are germane to the computation of oceanic and atmospheric flows comprise the field of computational geophysical fluid dynamics. It is a subdiscipline of the general field of computational fluid dynamics. However, com-

---

In the case of a marine ecosystem, the state variables could be the abundance of groups of species, phyto- and zooplankton, bacteria, nutrients, but they could also include key species. Different from purely physical systems, there is no consensus about the the state variables for ecosystems.

putational geophysical fluid dynamics has developed independently and in isolation from other fields of computational fluid dynamics such as aerodynamics. The main reason is that geophysical flows are more complex and less understood than aerodynamic flows and results are not as easily verified as aerodynamic computations are – by wind tunnel or controlled in-flight measurements. Also, much of aerodynamics is concerned with flows for which the Mach number, the ratio of fluid to sound speed, is order one or larger. Indeed one of the major aerodynamic problems is the accurate modeling of the supersonic shock structure without introducing unphysical oscillations and without smearing out the gradients. Geophysical flows on the other hand are low Mach number flows of a stratified, rotating, multicomponent fluid. The major challenge is the simultaneous computation of the many processes that affect such a flow rather than the highly accurate computation of a single process. For this reason computational geophysical fluid dynamics has developed on its own and has been somewhat conservative and slow in adopting the highly advanced algorithms of computational aerodynamics such as flow adaptive grids.

For many years, numerical problems were considered less relevant for tidal and climate studies, apart from the problem of designing a positive, monotonous and mass-conserving advection scheme (such as the semi-Lagrangian method). Recently, new challenges have emerged for efficient implementation of numerical codes on vector and parallel computers. It is expected that informatics and applied mathematics will play a more important role in environmental science in the future.

1.6 Applications

In providing a virtual reality, quasi-realistic computer models can be employed for various purposes. In particular, "experiments" impossible with the real system can be carried out. In the framework of fundamental sciences, such models allow the testing of hypotheses and extended simulations. Typical hypotheses concern the relevance of certain processes. Simulations generate complex data sets that allow detailed diagnostic studies of processes for which adequate observational evidence is lacking. In applied science, such models serve to interpret sparse and uncertain observational data, to forecast future states and to derive detailed scenarios of plausible future developments. Clearly, the separation into fundamental and applied sciences is blurred. In Chaps. 5 and 6 these applications are elaborated upon in some detail. Here, we want to give a first brief sketch of the spectrum of these applications.

Hypothesis testing

Experiments with computer models allow the formulation and test of hypotheses. Such experiments are called *numerical experiments*.

The aim is to understand and comprehend the system. The typical question is: what are the most important processes governing the system? Thus, such experiments are a tool in fundamental sciences. On the other hand, the impact of some human interference may also be studied in a numerical experiment. Then the effort is part of applied sciences. Some typical examples follow.

Example 1.1. A typical fundamental science question is the effect of cirrus clouds on climate. To answer this question three climate model simulations were performed which differed in the specification of cirrus clouds only [101]. In the first case, a standard parameterization was used, which gave a rather realistic vertical distribution of tropospheric and lower stratospheric temperatures. In the second simulation, the cirrus clouds were assumed to be black, and in a third one to be transparent. It turned out that black cirrus clouds lead to markedly increased tropospheric temperatures, by virtue of the greenhouse effect, with a concurrent cooling of the lower stratosphere. In the simulation with transparent cirrus clouds the effect was opposite. The conclusion drawn from these three experiment was that the radiative processes in cirrus clouds are first order processes. Without their proper inclusion, the simulation of the tropospheric and lower stratospheric temperature would deviate markedly from observed values.

Other problems can be addressed in a similar manner: the effect of vegetation, the release of large volumes of melt water into the Atlantic at the end of the last glacial period, or the barrier effect of the American isthmus on the global oceanic circulation.

Example 1.2. A problem in tidal modeling is the role of gravitational self-attraction of the ocean. This problem can be addressed by running a tide model twice, once with and once without self-attraction. The difference between the two simulations is indicative for the relative importance of the process "self-attraction" and tells the experimenter whether this process should be included in future simulations with this model or if the inclusion increases the complexity and computational load without having a significant effect on the results.

Example 1.3. The predictability of climate and its components is another problem that can be investigated with the help of two simulations of the same model. Initialize the model with two slightly different states. After a certain time the concurrent states of the two runs will differ like any two randomly selected states. Then the limit of predictability has been surpassed. This limit is a few days in case of mid-latitude weather and often much longer in the tropics and subtropics, due to the El Niño phenomenon.

Example 1.4. The effects of a bridge crossing the Øresund, between Denmark and Sweden, on the currents in the Øresund and on the flushing of the Baltic Sea have also been addressed by numerical experiments. Results are directly related to managerial decisions.

Some of these numerical experiments are also called sensitivity experiments since they determine how sensitive a model result is to changes in the value of a parameter or to the inclusion or omission of a process. For instance, one speaks of "the sensitivity of the mean atmospheric temperature to changes in the atmospheric composition".

Simulation of present and past states

The dynamics of actual and past regimes is of major interest for environmental sciences. What is the energy cycle of the atmosphere? How much eddy potential energy is transformed into eddy kinetic energy in cyclones? These questions can easily be answered by analyzing an atmospheric model, whereas the observations are currently not sufficient for such an analysis. Also the reconstruction of paleoclimatic states is a topical task in contemporary climate science. While some features may be deduced from various proxy data, such as isotope ratios in ice cores and width of tree rings, a spatially complete and dynamically consistent reconstruction can be made only with a climate model subjected to the appropriate forcing conditions such as land–sea distribution, atmospheric composition and orbital parameters.

Another application is the reconstruction of pathways and depositions of substances which have been emitted into the ocean or atmosphere. We will discuss in detail the case of lead after the burning of leaded gasoline was legally regulated in Europe (see Sect. 5.2.3). In this case, the spatially and temporally detailed reconstruction serves as a means to assess a posteriori the success of the regulation.

Simulation of likely future states

Apart from academic applications, there are a number of operational applications, most importantly the forecasts of weather[13]. Weather forecasts are part of everyday life. They go routinely into all kinds of planning activities. Spatial and temporal accuracy is important. Forecasts of the tides, ideally combined with forecasts of the wind effect on the water level, are also standard information that goes into operational decisions about ship traffic and other coastal activities. The forecast of extreme events, such as storm surges, is another example.

In all cases, the aim of the modeling effort is to get useful information about the *state* of the system. Attempts for a better understanding of the system are not made.

---

[13] The two terms *forecast* and *prediction* are often used interchangeably. In this book, however, we use the term "forecast" for efforts to specify an unknown state in the future, whereas the term "prediction" is used in a broader sense. It covers not only forecasts but also efforts to answer "what-if" questions, such as to predict the outcome of an experiment about the influence of a certain factor on the system.

Simulation of plausible future states

Predictions of the detailed future development of the atmosphere are limited to lead times of mostly a few days, at least in mid-latitudes. Society and stakeholders thus request a different type of prediction, namely the prediction of plausible future statistics of the atmosphere, conditional upon certain human activities. These conditioning human activities are almost never themselves predictable, so that the statistics of the atmosphere can also not be predicted. Instead, one refers to *scenarios*. When the assumptions about the human activity are plausible, then the scenario is plausible. When the assumptions are likely, the scenario is likely. When the assumptions describe an unlikely development, then the scenarios are unlikely.

Such scenarios are indeed the most intriguing applications of quasi-realistic climate models for the public. The best-known examples are the scenarios of future anthropogenic climate change, issued for example by the Intergovernmental Panel on Climate Change (IPCC) in its Assessment Reports. They describe the effect of the accumulation of carbon dioxide and other greenhouse gases and aerosols in the tropo- and stratosphere (see Sect. 5.3). Other scenarios study the effect of large-scale deforestation or the climatic impact of nuclear war.

Two common aspects of all these scenarios are that they are not forecasts and that they are usually not confirmed. The conditioning events might not take place at all, or the conditioning events may be changed by the availability of the scenarios, or effects might only appear in the far future. Unfortunately, scenarios often undergo a metamorphosis from something plausible or possible to something certain on their way from economic science to climate science to climate impact research to the public. A special problem emerges if conservative risk estimates are used in every step of the chain of assessments. Concatenating conservative estimates of risks sometimes results in unlikely worst case scenarios.

Examples of more regional or local scenarios include the effect of burning oil wells in Kuwait, the fate of waste heat from power plants, the impact of water quality regulations by the European Commission on the quality of dredging material in the port of Rotterdam, and the effect that the deepening of the shipping channel in the river Elbe may have on the tidal regime in Hamburg.

These simulations are a part of applied science. The typical question is how does the state of the system change if forcing, boundary or initial conditions are modified.

Data analysis

A relatively new application of quasi-realistic models is the "dynamically consistent interpolation" of irregularly distributed inaccurate observational data. In all the applications mentioned so far, the model is the primary source of information. It might have used observational evidence for the specification of boundary and initial conditions or parameterizations. But the knowledge gain comes from the information contained in the dynamical equations. In the case of the data analysis, the basic concept is different. The data are considered primary. The model is "only" used to fill the spatial and temporal gaps between the data "points"[14].

Data analysis is an application of a more general method called *data assimilation*, which is discussed in [141]. Data assimilation is based on a set of "dynamical equations" that approximately describe the evolution of the state variables and a set of "observation equations" that relate the observations approximately to the state variables. The unknown state – for instance a detailed weather map – is then estimated by optimally blending the observations with dynamical forecasts.

Such data analysis is routinely done in weather analysis and forecasting. Complete 3- or 4-dimensional synoptic representations of the atmosphere are constructed. They allow the study of processes which are not directly observable, such as the global meridional transports of heat and water. Data analysis is also applied, in a more experimental mode, to the El Niño and La Niña phenomenon, the state of the global ocean, the water quality in coastal waters, or the wave and current fields in the entrance area of ports.

Fig. 1.13 sketches the different purposes of various models, generating either practical knowledge to be used in social contexts or in generating dynamical insight to further scientific knowledge.

1.7 Issues

Quasi-realistic computer models are used extensively in environmental sciences, with the aim to generate new knowledge. But the models generate numbers, huge amounts of them, and not knowledge. Knowledge needs to be constructed from these numbers. An important first epistemological issue is thus how to turn these numbers into useful knowledge. A second epistemological issue is the basis for trusting the knowledge so generated. How can we be sure? Also, running quasi-realistic computer models is a human activity and, like any other human activity, subject to political and economic constraints and social and psychological conditioning. These are important issues and are dealt with in the respective sciences. Here we highlight some major aspects.

Reduction of information

Computer models generate huge amount of numbers. These numbers explicate the information contained in the dynamical equations and initial conditions. These numbers need to be reduced to useful information and knowledge. This

---

[14] In many cases the observations are in fact measurements taken at discrete points, but in some cases measurements also provide fields, as is the case for satellite retrievals.

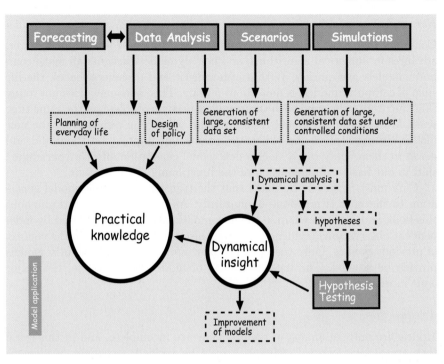

Fig. 1.13. Sketch of the different purposes of models, in generating either practical knowledge to be used in social contexts or dynamical insight to further scientific knowledge

requires a careful design of the numerical experiments and simulations so that proper reduction techniques indeed disentangle the complex output of the mathematical model. Often this reduction is accomplished by statistical methods, assuming that complex deterministic systems, such as the climate, behave in many respects like structured random systems.

In a fundamental science context, the ultimate goal of a model is often to understand a dynamical relationship, e.g., the stability of the Gulf Stream and its sensitivity to spatially changing fresh water forcing. Then reduced, i.e., cognitive models need to be fitted to the numbers. Care and ingenuity are required to identify, isolate and extract the relationship to be studied from all the other dynamical processes represented in the model.

In an applied science context, the purpose of the model is often to answer specific questions of a client. Such clients usually do not have a great deal of understanding of the dynamics but are merely interested in, say, the possible impact of climate change on agriculture. Then the information relevant to the client needs to be extracted.

These aspects are dealt with in Chaps. 5 and 6.

New and old models

Computer modeling is a rapidly developing field. Increase in computer power, advances in numerical algorithms, and new dynamical insights all make computer models age quickly. Within atmospheric and oceanic sciences, the life span of computer models is now about 5 years. This, of course, does not imply that the results obtained with these models become obsolete on the same time scale. Most of the old model results remain valid. The results from the new models do not replace the old model results but improve on them and add detail to them. Only rarely does a new model generation affect a "paradigm" shift in our basic comprehension of the functioning of the system.

Care must, however, be taken that the transition from one model generation to the next is not done haphazardly. As will be discussed throughout this book and again in Chap. 6 the new models need to be validated by independent observational evidence, but they also need to be put into the context of previous models. Only then can the new information that the new models provide be trusted. The trustworthiness of models is, of course, the major issue in modeling.

Trustworthiness

Models are only an analog of reality. They are incomplete, and in this sense they are always "wrong". Models cannot be verified. It cannot be demonstrated that they produce the right results for the right reasons. Models can only be validated. It can only be demonstrated that their results are consistent with observations. Models are validated by model-data comparison in parameter ranges for which observational data exist. Their usefulness, appeal and danger lie, however, in the option of applying them outside the validated parameter range, or, in Hesse's terminology, in the use of neutral analogs. To what extent can we trust such extrapolations? What is the basis for such trust? This trust does not only rest on hard sciences like physics but mostly on plausibility and other subjective assessments.

We deal with the problem of validation in more detail in Sect. 3.1.

Model builder and users

Quasi-realistic modeling in environmental sciences requires the cooperation of many scientists from different disciplines and a complex infrastructure. Environmental scientists work together with applied mathematicians and statisticians in large research centers that maintain dedicated computer centers, supported by hard- and software specialists. Computer modeling is "Big Science", with all the advantages and problems that such an approach entails. Among the advantages are the pooling of resources and the focus of activities. One major problem is that no single scientist comprehends all aspects of computer modeling any more, from the dynamics of the environmental system to the architecture of computers. A particular division has arisen between model builders and model users, where the model users do not necessarily understand the inner workings of the model and code any more. This is like driving a car without knowing how a combustion engine works, which is fine in most cases but can lead to disastrous consequences in certain situations. Other issues of big science are to justify and secure its substantial funding and to efficiently organize and administer such an enterprise. Quasi-realistic models are not a common tool. It is a rich man's tool. Certain groups and countries cannot afford it.

Social and psychological conditioning

Computer models and their virtual reality are created by the model builder and user. Whether a model is sensitive to changes in forcing and configuration is for the most part a matter of the system, but can in part be controlled by a skillful modeler. And there are many social and psychological mechanisms to induce the modeler to do just that, partly consciously and partly unconsciously.

There are no universal safeguards against such possible manipulation. Most manipulations are usually not a conscious act, but a social process, fueled by the wish of scientists to present results consistent with the current paradigm, to confirm plausible hypotheses and to promote funding and careers (see critique of the ENSO forecast community voiced by Landsea and Knaff [94]). In addition, certain properties of models are "socially favored", for instance high sensitivity. Journals like *Nature* and *Science* prefer "interesting" articles, i.e., disquieting or exciting scientific news. Also, attention by the general media helps funding and recognition of the researcher – and such attention is more easily obtained by alarming results than by reassuring results.

The classical peer-review process helps to avoid obvious excesses, but it also conservatively hampers the emergence of new concepts [38], [90]. Another safeguard is that modeling groups usually do not work independently but within a now global scientific, administrative and social network. This is especially true for groups working on scientific issues of general interest, such as the Global Warming problem. Scientists from all over the world interact, by direct cooperation, by meeting at conferences, and by reading each others publications. In the process they cross check their results. If the differences are too large, the reasons for it are identified, the models are corrected or modified, and a new more uniform set of results emerges. Nevertheless, this laudable scientific practice still leaves room for some subconscious collective biases. Thus consistency among models is not always a strong argument for the validity of the models, but may reflect a strong social convergence process among the model building institutions.

2

Computer Models

In this chapter we describe the basic elements that go into the construction of computer models and the salient properties of such models. The computer models in this book are based on dynamical equations that represent the laws of nature for the system under consideration. These dynamical laws are partial differential equations in space and time for a set of state variables. These equations can in most cases not be solved analytically but must be solved numerically. This requires first the discretization of the equations, both in space and time. The continuous problem with its infinite number of degrees of freedom is reduced to a discrete problem with a finite and manageable number of degrees of freedom. The discretized equations are then turned into a computer code that can be executed on a computer. This code then is the computer model.

The output of these computer models are numbers. These numbers must be interpreted and analyzed in order to derive useful knowledge. Formally, our models constitute dynamical systems, and concepts from dynamical systems theory are useful in this interpretation, especially when the same computer code is executed many times under slightly different (initial) conditions. In many cases, it is also useful to consider the model output as a realization of a stochastic process, though it is generated by a deterministic algorithm.

In Sect. 2.1 the process of deriving the governing differential equations is reviewed, which includes several closures through parameterizations and approximations. These equations are transformed into a discrete, finite form (Sect. 2.2), which allows for a digital implementation on a computer (Sect. 2.3). In Sects. 2.4 and 2.5 models are related to dynamical systems theory and to stochastic concepts. Finally, in Sect. 2.6 different types of forecast, exploiting initial conditions and external forcing, are discussed.

2.1 Dynamics

2.1.1 The Fundamental Laws

The dynamical laws for a fluid system such as the ocean or the atmosphere are the subject of fluid dynamics. First the state variables that describe the fluid must be specified. In the case of the ocean or atmosphere the state variables are usually the velocity, the pressure, the temperature and the salinity (humidity). Then the fundamental laws of physics, namely the conservation or balance laws for mass, momentum and energy are formulated in terms of these state variables. The resulting equations contain coefficients, such as the density and heat capacity that must be specified in terms of the state variables. These specifications rely on thermodynamic laws. The equations also contain terms that account for the fact that the ocean or atmosphere is not a continuous fluid but consists in reality of discrete molecules. These molecular effects are accounted for by phenomenological or molecular flux laws. The basic dynamical equations for a fluid thus consist of:

- the balance equations for mass, momentum, and energy,
- the thermodynamic specifications of the fluid, and
- the phenomenological flux laws.

These basic dynamical equations must be augmented by suitable boundary conditions. The equations and boundary conditions also contain external parameters, such as the gravitational acceleration or the rate of the earth's rotation that must be specified by the modeler.

These dynamical equations are derived and described in Appendix A.1. They are well established and experimentally proven. There is no problem with these basic laws. Problems arise when these laws are applied to real systems.

2.1.2 The Closure Problems

An environmental system consists of many interacting components that form a causal network. The causal links are dynamical processes. When one applies the laws of fluid dynamics to such a system one encounters three closure problems since it is impossible in a dynamical model

- to represent all processes within the system,
- to incorporate the surroundings, and
- to resolve all scales.

The **first closure problem** arises from the fact that more and more components and processes need to be incorporated for a complete description of the system. It becomes impossible to consider all components and processes, as is demonstrated by the following two examples.

Example 2.1. Cloud formation. Consider the atmospheric component of a climate model. At the most basic level the atmosphere is treated as a one-component system. It simply consists of air. One only needs to consider a single mass balance equation. At the next level one takes into account that air actually consists of dry air and water vapor. One treats the atmosphere as a two-component system with two mass balance equations, one for dry air and one for water vapor. However, this description is still not complete since water vapor can turn into clouds which consist of water drops or ice crystals. These phase changes are accompanied by release or gain of latent heat. To include the formation of clouds into the dynamical description of the atmosphere one must consider three balance equations for water: one for each phase. These balance equations must include source and sink terms that account for the condensation/evaporation, deposition/sublimation and melting/freezing of water. One also needs to add to the internal energy equation terms that account for the latent heat released or required for these phase changes. Appropriate forms of these balance equations can be found in Appendix A.5.1.

However, this is not sufficient. Upon closer examination one finds that the formation of cloud drops or ice crystals is a fairly complicated process. It consists of two stages. A nucleation stage and a growth stage. Formation of clouds requires a supersaturated atmosphere for which the water vapor pressure is larger than its equilibrium value. Cloud drops (or ice crystals) then form spontaneously. Drops of small radii form more likely than of large radii. However, for a drop to be stable the latent heat released by condensation must exceed the surface tension work required to form the drop. This is only the case if the drop radius is larger than a certain critical radius. This critical radius decreases with increasing supersaturation. For typical supersaturations occurring in the atmosphere the critical radius is fairly large and spontaneous nucleation hence very rare. This is true in a "clean" atmosphere. However, the atmosphere contains hygroscopic aerosol particles such as $NaCl$ or $(NH_4)_2SO_4$. Condensation at these particles lowers the critical radius since the resulting solution has a lower equilibrium pressure than pure water. The atmosphere appears more supersaturated. Cloud drops are primarily formed from these aerosol particles, the cloud condensation nuclei. To correctly model cloud formation one thus needs to know the distribution of cloud condensation nuclei. These are formed naturally (like sea salt from breaking waves or dust from volcanic eruptions) or anthropogenically (like SO_4 by burning of fossil fuels). The issue is further complicated by the fact that the critical radius depends on the size of the aerosol particle. We not only need the number of aerosol particles in a certain volume but also their size distribution.

After nucleation, cloud drops grow by diffusion of water vapor towards the drop. The latent heat released by condensation diffuses away from the drop. These two processes are described by the phenomenological flux laws. It is found that the growth rate of cloud drops depends on the drop size. One thus needs the size distribution of cloud drops. Further growth of cloud drops is enhanced by collision and coalescence until rain drops are formed that precipitate.

Example 2.2. Radiation. Similar considerations apply to the energy balance of an atmospheric model. There are different forms of energy: mechanical, internal and radiative energy, among others. Energy can be transformed from one form into another. Radiative energy is converted to internal energy when radiation is absorbed by matter. Internal energy is converted to radiative energy when matter emits radiation. These absorption and emission processes depend on the wavelength of the radiation. One must distinguish between short-wave or solar radiation and long-wave or thermal radiation. To model these processes one needs separate balance equations for internal and radiative energy. The radiative energy equation must further be split up into one for solar radiation and one for thermal radiation. Explicit forms of the radiative energy equation can be found in Appendix A.5.2.

The absorption and emission processes also depend on the concentration of various radiatively active constituents in the atmosphere, such as water vapor, clouds, carbon dioxide and ozone. One thus needs balance equations for all the radiatively active constituents. Absorption and emission might also be accompanied by chemical reactions. These photochemical reactions are discussed in Appendix A.5.3.

These two examples are not academic but at the core of climate modeling. Climate is determined by the balance of the incoming solar radiation and the outgoing thermal radiation (cf. Fig. 1.11). The cloud cover determines how much of the incoming solar radiation is reflected. As described above, the details of these radiative and cloud formation processes depend in a complicated manner on the wavelength of the radiation and on the concentrations of radiatively active gases, aerosols and dust particles, which in turn depend on processes at the air–sea interface, volcanic eruptions, human activities, and other processes. As one might imagine there is no end to this causal network. The closer one looks, the more components and processes must be included.

There is no obvious point where to cut this causal network. This problem constitutes the first closure problem. It arises because the environmental system is a complex system. It consists of an unmanageably large number of interacting components.

A **second closure problem** arises because the system is open. It interacts with its surroundings. The system exchanges mass, momentum and energy with its surroundings. This exchange must obey the boundary conditions in Appendix A.4 which state that certain fluxes and variables have to be continuous across the bounding interface. These continuity conditions are useful for the prediction of the system only if a sufficient number of fluxes and variables are prescribed at the interface. Thus models of the ocean assume prescribed values of the momentum, heat and fresh water flux at the air–sea interface. These prescribed boundary conditions represent the effect of the surroundings on the system, the effect of the atmosphere on the ocean.

However, we can never prescribe these boundary conditions exactly. One reason is that the surrounding is affected by the evolution of the system. This is obviously true in the above example. The evolution of the ocean affects the atmospheric circulation, which in turn affects the values of the fluxes at the air–sea interface. The specification of the boundary conditions represents our second closure problem.

The specification of external parameters represents a related closure problem. External parameters such as the rate of rotation of the earth and the geopotential in the momentum balance are not truly external parameters. The gravitational potential is determined by the distribution of mass, and mass is redistributed by ocean currents and atmospheric winds. Gravitational self-attraction has indeed become an issue in tidal modeling. Similarly, the rate of rotation of the earth also changes because of mass redistribution (affecting the moment of inertia) and angular momentum exchanges within the earth–moon system. There are no truly external parameters. Specification of parameters represents a closure problem. The decision of what can safely be treated as an external field or parameter depends on the goal of the modeling effort, the time scales of the involved processes and other contingent factors and requires considerable insight into the functioning of the system.

A **third closure** problem arises because one cannot resolve processes at arbitrarily small temporal and spatial scales. One can only manage a coarse-grained or discrete description. This might not look like much of a problem since one is usually only interested in such coarse information. An average wind speed of 10 m/s in the North Sea is valuable information for estimating the height of the waves in that area, but the wind may be considerably larger during spatially and temporally limited gust events. These small-scale features have a significant effect on both wave generation and the mixing of the upper ocean. They also have a significant effect on the wind stress. Formally, the flux of momentum across the boundary is given by

$$\mathbf{F} = C_D \mathbf{u}^2$$

where C_D is the drag coefficient and $\mathbf{u}$ the air velocity (See Appendix A.8.1). The air velocity consist of a mean value, indicated by a bar, and a fluctuation about this mean, indicated by a prime

$$\mathbf{u} = \bar{\mathbf{u}} + \mathbf{u}'$$

This decomposition is called the *Reynolds decomposition*. The mean momentum flux across a boundary thus consists of two parts $\bar{\mathbf{F}} = C_D \bar{\mathbf{u}}^2 + C_D \overline{\mathbf{u}'^2}$. The first term describes the flux of momentum due to the mean wind and the second term the wind stress exerted by the fluctuating velocity. The important point is that the second term, $\overline{\mathbf{u}'^2}$, does not vanish. There exists a mean momentum flux

$$\mathbf{F}_{edddy} = C_D \overline{\mathbf{u}'^2}$$

which is caused by the unresolved fluctuating components. This flux is called the *subgridscale* or *eddy*, or *Reynolds flux*. Such fluxes also occur for other quantities such as heat and salinity or humidity. The details are discussed in Appendix A.6. The important point is that these fluxes affect the resolved scales but are determined by the unresolved scales. One needs to know the subgridscales to calculate the coarse-grained fields. This constitutes a closure problem. It is a manifestation of the advective and other nonlinearities in the dynamical equations which couple scales.

2.1.3 Parameterizations

The need for parameterizations arises because of the closure problems. When one cuts the causal network at any point, one has to account for the effect of the components and processes that are being cut off on the components that are explicitly calculated. When a certain part of the environment is singled out, one has to account for the effect of the surroundings. When a certain coarse graining is adopted one has to account for the effect of the unresolved processes on the resolved ones. Parameterization is the specification of these effects in terms of the resolved variables or "external" parameters. However, there is no way to accurately account for these effects other than by actually including the processes that are being excluded, adding the surroundings to the system that is being modeled and resolving the unresolved scales. All parameterizations of these effects are just patch-up jobs. It is not clear whether these effects can be expressed in terms of the resolved variables or prescribed as external parameters. Parameterizations are most often educated guesses, approximative at best. Parameterizations thus introduce errors into the dynamical equations and hence into the model.

The statement that parameterizations introduce errors is obvious for a climate model that prescribes the cloud cover and the surface albedo rather than calculating them. It is also true for the parameterization of subgridscale fluxes. Since these fluxes arise from coarse graining in much the same way as molecular fluxes, they are often modeled mimicking the molecular or phenomenological flux laws. Eddy diffusion coefficients are introduced instead of molecular diffusion coefficients. These eddy diffusion coefficients are, however, a property of the subgridscale flow, not a property of the fluid. They need to be readjusted if the resolution changes. We are not on solid ground any more. Parameterization of eddy fluxes are not on a par with the phenomenological flux laws.

Climate and other modelers try to alleviate the closure problems by including more and more components and processes into their models, by considering larger and larger systems, and by increasing the resolution of their models. Thus processes that need to be specified become resolved. However, these efforts only push the boundary. They do not solve the closure problem. The climate system has an infinite number of interacting degrees of freedom. A climate model can only explicitly calculate the evolution of N degrees of freedom. The effect of the unresolved degrees of freedom must be prescribed however large N is.

As one resolves more and more processes the character of the dynamical equations changes. This change of structure has an interesting implication for the mathematician. When asked by the mathematician: "What are the governing equations?", the modeler has to reply with the question: "At what resolution?". The character of the equations may change abruptly when the grid spacing Δx is changed. Thus, the limiting process $\Delta x \to 0$ is not properly defined and standard concepts in numerical mathematics, like "consistency" and "convergence" are not applicable.

2.1.4 Approximations and Representations

In contrast to parameterizations, approximations are applied to the resolved dynamics in order to eliminate processes and aspects that are considered minor or irrelevant. A typical approximation for the atmosphere and ocean is the shallow water approximation, which is described in Appendix A.9.2. Other approximations are the planetary and quasi-geostrophic approximations. These approximations, their validity, and their applicability are the topics of textbooks. In our context, they represent a more technical aspect.

Dynamical equations can also be formulated in different coordinate systems and by using different but equivalent sets of dependent variables. One might use height or isopycnal coordinates. Instead of velocity one might use vorticity and divergence. Such different representations are described in Appendix A.10.1. Different choices have different properties and some choices are better suited for certain problems than others.

In summary, models are based on the integration of a set of approximate dynamical equation. The word "set" means that only a limited number of state variables $\psi(\mathbf{x}, t)$ are chosen to describe the system. In the case of tides, these are usually the surface elevation (or water depth) and the horizontal volume transport (or depth-integrated horizontal velocity). In the case of the ocean or atmosphere the state variables are usually the velocity, the pressure, the temperature and the salinity (humidity). These state variables are functions of position $\mathbf{x}$ and time t. Once the state variables are chosen one makes appropriate approximations to the basic dynamical laws, parameterizes subgridscale and other unresolved processes, specifies external parameters and chooses a suitable representation. One then arrives at a set of dynamical equations for the state variables. These equations are partial differential equations in space and time. They consist of a set of prognostic equations that govern the time evolution

$$\partial_t \psi(\mathbf{x}, t) = \mathcal{A}[\psi(\mathbf{x}, t); \mathbf{x}, t; \tilde{\boldsymbol{\alpha}}(\mathbf{x}, t)] \tag{2.1}$$

and a set of diagnostic equations that relate different variables

$$\mathcal{B}[\psi(\mathbf{x}, t); \mathbf{x}, t; \tilde{\boldsymbol{\beta}}(\mathbf{x}, t)] = 0 \tag{2.2}$$

Here $\mathcal{A}$ and $\mathcal{B}$ are differential operators in space and $\tilde{\alpha}(\mathbf{x}, t)$ and $\tilde{\beta}(\mathbf{x}, t)$ sets of parameters that describe the external fields and parameterizations. These differential equations have to be augmented by appropriate initial and boundary conditions. It is these equations that the dynamicist hands down to the numerical mathematician. They are valid only for certain parameter ranges, and – uncommon for mathematicians – only for certain space/time scales.

2.2 Numerics

The dynamical equations of fluid flows are a set of coupled nonlinear partial differential equations (PDEs). These PDEs can be solved analytically only for highly idealized cases. Solutions for even modestly realistic situations must be obtained numerically.

When a PDE is solved numerically the continuous functions are represented by their values at preselected discrete points in space and time. Derivatives are replaced by finite differences. This is the standard grid method. Alternatively, the continuous functions are expanded into a finite truncated series of basis functions. The differential operators then act on the basis functions with known results. Hybrids of these method also exist. All these numerical methods reduce the continuous problem with its infinite number of degrees of freedom to a discrete problem with a finite and manageable number of degrees of freedom. This reduction of course introduces errors, truncation errors. Numerical methods must be designed to keep this truncation under control. The method must be stable. Otherwise the numerical solution "explodes" and becomes useless.

As an example consider the linear differential equation

$$\frac{du}{dt} = -\gamma u$$

with initial value $u(t = 0) = u_0$ and $\gamma > 0$. It has the solution $u(t) = u_0 e^{-\gamma t}$, which tends to zero for $t \to \infty$. Using the forward difference or explicit scheme, we get

$$\frac{u_{n+1} - u_n}{\Delta t} = -\gamma u_n$$

or

$$u_{n+1} = (1 - \gamma \Delta t)u_n$$

For $\Delta t > 2/\gamma$ the value $|u_n|$ increases monotonically with n and the solution is unstable. A sufficiently small time step is required for stability.

Other important aspects of numerical schemes are their accuracy and efficiency. These and other more technical aspects like consistency, convergence, forward, backward, and centered differences, and explicit and implicit schemes are discussed in Appendix B.2[1]. The details depend on whether the underlying PDE is elliptic, parabolic or hyperbolic.

When constructing numerical algorithms for atmospheric and oceanic flows new issues come into play because of the specific nature of these geophysical flows. First of all, geophysical flows are very anisotropic[2]. The vertical direction is distinctively different from the horizontal directions. Indeed, as will be discussed in Appendix A.10, it is often advantageous to replace the vertical coordinate by a dynamical variable such as the (potential) density, pressure or entropy and introduce isopycnal, isobaric or isentropic coordinate systems. Discretization is then done in these new coordinate systems. So the finite difference grid is not necessarily a regular grid in physical space. A second issue is that the equations describing geophysical flows contain more than one dependent variable. This leads to the possibility of "staggered" grids[3]. As mentioned above, finite difference grid methods are not the only way to discretize the equations. The spherical geometry of the earth makes atmospheric flows amenable to spectral methods where the variables are expressed as a truncated sum of spherical harmonics. The complex shape of the ocean basins, on the other hand, makes finite element methods attractive for ocean modeling. Spectral element methods try to combine the advantages of spectral and finite element models. These methods are discussed in Appendices B.4 and B.5.

Formally, the numerical discretization is usually done in two steps. The first step is the discretization in space, by either using the grid, spectral or finite element method. At any instant of time the state of the system is then described by the state vector, denoted by $\boldsymbol{\psi}(t)$. It is the collection of all state variables at all grid points, or the spectral expansion coefficients of all state variables, or the coefficients of all state variables in a finite element expansion. The dimension of this state vector is the number of state variables times the number of grid points, or basis functions, or finite elements. The partial differential equation (2.1) becomes an ordinary differential equation (ODE)

$$\frac{d}{dt}\boldsymbol{\psi}(t) = \mathbf{A}[\boldsymbol{\psi}(\mathbf{t}); \mathbf{t}; \boldsymbol{\alpha}(\mathbf{t})] \tag{2.3}$$

with an algebraic operator $\mathbf{A}$. The diagnostic equations and boundary conditions become absorbed into the operator $\mathbf{A}$ and parameter $\boldsymbol{\alpha}(t)$. We will use this equation as a prototype for our discussion. Final discretization in time leads to a discrete map

---

[1] Useful introductions into the numerical treatment of differential equations, with particular emphasis on the needs in atmospheric and oceanographic modeling are provided by Arakawa [1], Arakawa and Lamb [2], Haidvogel and Beckmann [51], Haltiner and Williams [54], Kantha and Clayson [76], Mesinger and Arakawa [113], and Washington and Parkinson [186].

[2] A flow is called *anisotropic* if its characteristics depend on its direction.

[3] A grid is staggered when separate grids are used for different variables. For further details refer to Appendix B.3. An example is shown in Fig. 4.5.

$$\boldsymbol{\psi}_{i+1} = \mathbf{A}_i[\boldsymbol{\psi}_i; \boldsymbol{\alpha}_i] \tag{2.4}$$

where i is the discrete time index. This discrete map is then turned into a computer code. This computer code is the final product. It calculates $\boldsymbol{\psi}_i$ for $i = 1, 2, \ldots$, given $\boldsymbol{\psi}_0$ and $\boldsymbol{\alpha}_i$ for $i = 0, 1, \ldots$.

Stability, accuracy and efficiency are the most important properties of numerical algorithms. Stability requires a small time step and hence a large number of operations. Accuracy can be increased by increasing the resolution but this implies a larger number of operations and hence decreases the efficiency. Trade-offs must be made which are ultimately determined by the available computer resources.

2.3 Computers

The performance of digital computers has increased dramatically since their invention in the late fifties of the last century.[4] The clock speed of the central processing unit (CPU) is now[5] below a nanosecond, thus allowing a single processor to perform several gigaflops, i.e., about 10^9 to 10^{10} floating point operations per second. One can now store several gigabyte in core memory. Disc storage has also increased to thousands of gigabytes. These increases are indeed dramatic and now allow calculations which were not feasible just a few years ago. However, despite all this progress there are limitations.[6]

One limitation is that digital computers have a finite word length. Current computers use 32 or 64 bit words, which allows representing a floating point number to about 9 or 15 significant decimal digits. Because of this limited machine accuracy, floating point operations are only approximate and contain a round-off error. These round-off errors may accumulate and are a principal limitation of the accuracy of numerical calculations on digital computers. Furthermore, due to the round-off errors, identical instructions may produce slightly different results on different machines or even when compiled with different compilers. Stability of the numerical algorithms used is therefore very important. Decreasing the truncation error of a numerical algorithm beyond a certain point becomes counterproductive because the increase in round-off errors exceeds the decrease in truncation error. The machine accuracy can be doubled by software instructions but at the expense of the speed of the calculation.

The second major limitation of digital computers is the speed of their central processing units, which might come to their physical limits in the future.

---

[4] In writing this section, we were competently helped by Dr. Joachim Biercamp from the German Climate Computer Center in Hamburg.

[5] in 2003

[6] In the past 30 years, the computational efficiency of microprocessors has doubled every 18 months, on average – at a stationary price level. Specialists expect that this regular increase, named *Moore's law*, will continue, at least until about 2010 [49].

To increase the performance with constant clock speed one has developed vector processors. Vector processors can process several operations concurrently in one clock cycle, thus they are more effective but also much more expensive then traditional scalar processors. Another way to increase performance is to use more than one processor. Modern supercomputers use hundreds of vector processors and so called massively parallel processing systems may use thousands of cheaper scalar processors simultaneously. The theoretical peak performance of today's most powerful systems thus reaches several teraflops. However, the efficiency gained from such machines depends on the ability of a code to adapt to such machines. It is not always easy, sometimes even impossible, to vectorize given algorithms and to distribute them efficiently onto different processors. Consequently the gap between the theoretically achievable peak performance of a computer and the effective sustained performance which can be reached with a complex numerical model has become wider in recent years[7]. Writing an efficient code is nowadays a matter of vectorizing and/or parallelizing the code. It requires detailed consideration of the architecture of the specific computer on which the code is to be run.

Another relevant issue is portability. A model's lifetime is longer than that of a computer and in a collaborative scientific environment it is often desired to run it on more then one platform. Thus there is a tradeoff between efficiency on a given machine and the possibility to run the code on different machines.

Another contemporary problem with running complex models on fast computers is the problem of efficiently and timely storing the results of the calculations. This transfer of data from the central processing unit to another computer, responsible for storing and administrating large amounts of data, is far from simple and requires significant investments in hardware and software. An even greater problem arises in retrieval and processing of data. Data which have been written step by step during a simulation and are stored in hundreds and thousands of big files may be needed all at once to obtain information on the temporal evolution of a simulated system.

The finite computer power is the limiting factor of what problems can or cannot be done. The modeling community thus demands, and correctly so, larger and faster computers. But it must be kept in mind that modeling is not just a computational problem. Because of finite computer power, one can only solve equations that contain a limited number of physical processes. It requires considerable insight into the dynamics of a given problem to determine which processes can be neglected or parameterized and which processes must explicitly be calculated, and to formulate the corresponding set of equations. This is the task of the (fluid) dynamicist who is still at the basis of the modeling enterprise.

---

[7] Moore's law holds for peak performance only. The increase in sustained performance has been slower, especially for earth system modeling codes, that combine various components, such as the atmosphere, the ocean, the vegetation and the cycles of matter into one model.

When more computer power becomes available, two options for the modelers arise. One option is to increase the lengths and number of integrations, for instance by generating ensemble simulations. The other option is to increase the resolution and to include new processes, for instance by introducing atmospheric chemistry into climate models. The former task is straightforward, but the latter task requires considerable input from the dynamicist. Formulations and parameterizations for the new processes are needed. The existing equations and parameterizations need to be reformulated to be consistent with the increased resolution (cf. Sect. 2.1.3).

2.4 Models as Dynamical Systems

Computer models are based on discretized dynamical equations that are sets of ordinary differential equations (ODEs) or discrete maps, such as equations (2.3) and (2.4) above. These equations are examples of dynamical systems. For discussion purposes consider the autonomous system

$$\frac{d}{dt}\boldsymbol{\psi}(t) = \mathbf{A}[\boldsymbol{\psi}(\mathbf{t}); \boldsymbol{\alpha}] \tag{2.5}$$

where the algebraic operator $\mathbf{A}$ and the parameter $\boldsymbol{\alpha}$ do not depend on time. The solution of this equation depends on the initial condition $\boldsymbol{\psi}(t=0) = \boldsymbol{\psi}_0$ and on the parameter $\boldsymbol{\alpha}$. Thus

$$\boldsymbol{\psi}(t) = \mathcal{F}[t, \boldsymbol{\psi}_0, \boldsymbol{\alpha}]$$

For a specific initial condition and parameter value the solution is a trajectory in the N-dimensional phase spanned by $\boldsymbol{\psi} = (\psi_1, \ldots, \psi_N)$

$$t \to \mathcal{F}[t; \boldsymbol{\psi}_0, \boldsymbol{\alpha}]$$

If solutions for different initial conditions are considered, $\mathcal{F}$ becomes a time and parameter-dependent mapping of the phase space into itself

$$\boldsymbol{\psi}_0 \to \mathcal{F}[\boldsymbol{\psi}_0; t.\boldsymbol{\alpha}]$$

very much like a fluid flow. Dynamical systems theory studies the geometric and topological properties of this mapping, rather than the properties of a single trajectory. The properties of the mapping generally depend on the parameter $\boldsymbol{\alpha}$. All definitions, concepts and results of dynamical systems theory are about this mapping. The mapping is called conservative if it conserves the phase space volume, and dissipative if it does not. A subspace is called an attractor if all trajectories converge onto this subspace. Such an attractor can have a very complex structure, with any dimension between 0 and N, including fractal dimensions[8]. A point $\boldsymbol{\psi}_*$ that satisfies $\mathbf{A}[\boldsymbol{\psi}_*] = \mathbf{0}$ is called a

---

[8] The *fractal dimension d* of an object in a multi-dimensional space is the exponent of increase of mass (or any other property) when the radius of the object is increased. See also [16].

fix point. Fix points can be stable and act as attractors for their neighborhoods or be unstable.

The flow in phase space might be quite complex. A given finite volume element might be stretched out into ever finer filaments until the whole phase space is covered. A dynamical system is said to be chaotic when points in phase space that are initially close together diverge from each other at a sufficiently fast rate. The formal definition requires at least one of the Lyapunov exponents[9] to be positive. Such chaotic systems have limited predictability (i.e., become unpredictable beyond a certain time lag, see Sect. 2.6) since initial conditions are never known with infinite precision. In general, the character of the flow in phase space is related to the spectrum[10] of the operator $\mathbf{A}$.

These and other concepts and results are covered in textbooks such as [16] and [187].

Dynamical systems are deterministic. Nevertheless, they can exhibit quite irregular behavior, due to the nonlinearities of the system. Irregular behavior can thus be generated internally. It does not need to be imposed externally by irregular geometry or forcing or by randomness. This was a major insight.

Concepts from dynamical systems theory have been very useful for low dimensional systems. A famous example is the Lorenz system [102] with a dimension $N = 3$, which is supposed to describe features of convective systems. Fix points can be determined. Their stability as a function of control parameters can be determined. Numerical experiments can be performed to map out the attractor.

The application of dynamical systems concepts to high-dimensional systems, such as a quasi-realistic climate model, is more problematic. The determination of the spectrum of the operator $\mathbf{A}$ is computationally not within reach; and neither is the systematic exploration of the complete phase and parameter space. One tries to get by, by studying just one trajectory in phase space, or a few, but without any strict and solid results. Nevertheless, there is the consensus that many of these higher dimensional systems are chaotic, in particular systems that are turbulent, exhibit instabilities and are strongly affected by phase transitions. Many atmospheric phenomena, such as cyclogenesis and convective rainfall, fall into this category. Nevertheless, typical dynamical systems properties are not always found in these systems. A particular point in question is the emergence of multiple equilibria in chaotic systems. This phenomenon is "interesting" and hence studied extensively – with low-dimensional models. However, large-scale multiple equilibria have not yet been detected in the instrumental observational record of the atmosphere (see Sect. 6.3.2) nor in long-term simulations with atmospheric models.

---

[9] Lyapunov exponents are the eigenvalues of the matrix $\mathbf{A}^*$, obtained through linearization of $\mathbf{A}$ at the point ψ in (2.5). A positive Lyapunov exponent indicates that an initial disturbance grows. The Lyapunov exponent varies in phase space. In parts of the phase space the system may be chaotic, in others not.

[10] The spectrum of $\mathbf{A}$ at ψ is the set of eigenvalues of the linearized operator $\mathbf{A}^*$ at ψ.

Oceanic models (see Sect. 6.1.4) and paleoclimatic data, on the other hand, point to multiple equilibria and hysteresis effects.

2.5 Models as Stochastic Systems

There are two reasons to introduce randomness and statistics into environmental modeling:

- model output can often not be distinguished from random behavior, and
- the model equations are inexact.

As to the first reason, deterministic but chaotic dynamical systems often show an irregular behavior that cannot be distinguished from stochastic or random behavior. This fact is demonstrated in Fig. 2.1. It shows two time series. One represents the sum of 20 chaotic but purely deterministic processes. The other series is a series of random numbers which are realizations of a normal random variable, whose first two moments match those of the deterministic time series. While the two series are different at any time instant, their overall characteristics are very similar. The histogram of the deterministic time series is indeed close to a normal distribution, as a consequence of the Central Limit Theorem[11].

When one accepts this similarity, the solution of the model equations can be interpreted as a realization of a random or stochastic process. It dos not matter whether or not the process is really stochastic, as long as it cannot be distinguished from the output of a stochastic process[12]. The *mathematical construct* of randomness is just a convenient and efficient tool to bring order into seemingly irregular and unstructured data[13].

Meteorological observations and climatic variables are in fact sufficiently irregular that they are efficiently described by random distributions (e.g., [182]), in terms of extreme events, the range of variations, or temporal and spatial correlations. On the other hand, attempts to identify distinct nonlinear deterministic structures in the data sets fail in almost all cases. Therefore it has become common in meteorology and climate science to consider at least the instrumental observational record as the result of a stochastic process, which

---

[11] The similarity is usually rationalized by asserting that the evolution of the deterministic system is caused by independent impacts and equating independent impacts with random impacts. Randomness models independence. The exact meaning of these statements is pursued in the foundations of complex system theories.

[12] The old philosophical question of whether or not "God plays dice" may have a similar answer. God's actions may not be understood by us humans and appear as random but this does not exclude the possibility that God actually acts according to a plan.

[13] The adequacy of conceptualizing randomness as the highly irregular outcome of a deterministic process is underlined by the fact that random number generators on computers are specific deterministic formulae.

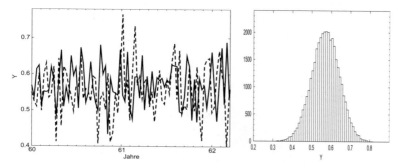

Fig. 2.1. Sum of 20 chaotic deterministic processes and a series of random numbers, whose first two moments match those of the time series of the sum (*left*). Frequency distribution of the sum of the 20 chaotic processes (*right*). From von Storch et al. [177]

is conditioned upon a variety of external and internal parameters. An important implication is that the identification of "signals" requires the discrimination between internal variations and the effect of, say, changed parameters or boundary conditions.

The tides are a much more regular deterministic systems and "signals" are easily detected, without the use of statistical tests or other statistical techniques. However, as soon as a system is influenced by weather, mostly as a forcing factor, it inherits the inherent randomness of weather. Examples are storm surges.

The second reason is that the dynamical equations are inexact. First they have undergone considerable approximations (Sect. 2.1.4), such as the elimination of sound waves in ocean and atmospheric modeling, and numerical manipulations (Sect. 2.2), such as the replacement of differential operators by difference operators. Second, the parameters, forcing functions, initial and boundary conditions are not exactly known. To account for the inexactness of parameterizations sometimes the concept of *randomized parameterizations* is invoked [173], [98], [97]. Random "noise" is introduced into the formulation of certain parameterizations, according to the Bayesian credo that every uncertainty should be modeled as a random process. Then, the dynamical equations become a set of stochastic differential equations. Equation (2.3) becomes

$$\frac{d}{dt}\boldsymbol{\psi}(t) = \boldsymbol{\lambda}(t)\mathbf{A}[\boldsymbol{\psi}(\mathbf{t}); \mathbf{t}; \boldsymbol{\alpha}(\mathbf{t})] + \boldsymbol{\xi}(\mathbf{t}) \qquad (2.6)$$

where $\boldsymbol{\lambda}(t)$ and $\boldsymbol{\xi}(t)$ are random processes, introducing multiplicative and additive "noise" to the equation. The solution $\boldsymbol{\psi}(t)$ now becomes a random function.

There are numerous examples that such a replacement of a deterministic model by a random model is indeed a good strategy. Consider the drag of the sea surface on atmospheric flows. It is parameterized by a drag law with

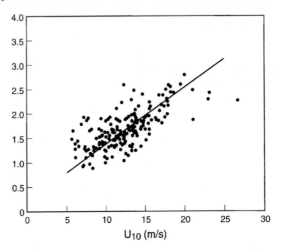

Fig. 2.2. Scatter diagram of simultaneous measurements of the drag coefficient C_D and of the wind speed U at 10 m height. The *straight line* is a regression line. After De Cosmo et al. [24] (© (1996) American Geophysical Union. Reproduced by permission of American Geophysical Union)

a wind-dependent drag coefficient, but observations (see Fig. 2.2) show that the drag coefficient actually varies in an apparently random manner. Part of the variations may be due to measurement problems. Part of the variations might be eliminated by including dependencies on other resolved variables, not just the wind speed. But a significant part of the variations will be related to unaccountable, local features that can only summarily be described as random noise.

Stochastic dynamical systems are often called complex systems. Their basic properties are described in textbooks (e.g., [64]). The basic probabilistic and statistical concepts that we will employ in this book, such as random variables, distributions, correlation functions and spectra, are briefly described in Appendix C. A more complete account of statistics in climate sciences is offered by von Storch and Zwiers [182].

2.6 Predictability

An important application of models is to forecast (or to predict) and predictability (or more accurately its limit) becomes an important issue. Predictability arises from both the knowledge of the initial condition and the dynamics. Consequently one has to distinguish between two different types of forecasts.

2.6.1 Limit of Predictability

In chaotic systems the evolution depends sensitively on the initial conditions. Differences in initial conditions grow exponentially. Though the evolution is deterministic it is predictable only for a limited time[14]. This time is called the *limit of predictability*. This time is usually estimated by considering the correlation function

$$\rho(\tau) = \lim_{T \to \infty} \frac{1}{T} \int_0^T dt \; \psi(t)\psi(t+\tau)$$

The time τ_* where $\rho(\tau)$ reaches zero is considered to be the limit of predictability. After that time, the state of the systems no longer depends in any identifiable systematic manner on the initial state. This does not imply that the state at time t is independent of the initial state at time $t = 0$. Different initial states will lead to different states at later times, but the character of this link becomes undeterminable. In case of mid-latitude weather this limit of predictability is of the order of many days; in case of El Niño it is many months, possibly even a few years.

Note that this definition of the limit of predictability uses a statistical concept, the correlation function. More properly, dynamical systems concepts should be invoked, but cannot for high-dimensional systems, as discussed in Sect. 2.4.

How does the limit of predictability enter the solution and affect forecasts? The evolution equations (2.3) can be integrated forward in time to determine the future state of the system. This calculated future state depends both on the initial condition ψ_0 and on the parameter vector $\alpha(t)$.

As an example consider the evolution equation

$$\frac{d}{dt}\psi(t) = r\psi(t) + \eta(t)$$

which describes a forced oscillator. It has the solution

$$\psi(t) = \psi(t_0)e^{r(t-t_0)} + \int_{t_0}^t d\tau \; G(\tau - t)\eta(\tau) \tag{2.7}$$

with the Green's or influence function

$$G(t) = e^{rt}$$

The behavior of this solution depends critically on the real part of the coefficient r. For $Re[r] < 0$ the oscillator is damped. The influence of the initial

---

[14] Note that chaotic behavior does not imply catastrophic behavior. A chaotic system has neither to move towards totally new states nor to change quickly and dramatically. Instead the system moves towards known states. Which states the systems moves to, however, depends very sensitively on the initial conditions.

conditions dies away and for large times the solution is solely determined by the forcing. For $Re[r] = 0$ the influence of the initial conditions persists. In the case $Re[r] > 0$ the oscillator is unstable. The influence of the initial conditions becomes dominant.

More general systems have characteristics similar to our simple example and their solutions may formally be written

$$\psi(t) = \mathcal{F}[\psi(t_0), t - t_0] + \int_{t_0}^{t} d\tau \; \mathcal{G}(\tau - t)\eta(\tau) \tag{2.8}$$

where the operator $\mathcal{F}$ and the Green's function operator $\mathcal{G}$ represent the internal dynamics and are independent of the forcing η. One must now distinguish two kinds of forecasts.

2.6.2 Forecast of the First Kind

There are systems for which the initial conditions become unimportant. "Initial transients" die away and the solution is essentially determined by the external forcing

$$\psi(t) = \int_{-t_0}^{t} d\tau \; \mathcal{G}(\tau - t)\eta(\tau) \tag{2.9}$$

Such systems are in principle predictable for infinite times, as long as the external forcing η is predictable. An example are the tides, as long as the geological configuration of the earth and the internal dynamics $\mathcal{G}$ remain unchanged, and the external forcing η remains known. In the case of the global tides η is the tidal potential of the moon and sun. In the case of the tidal inlet Jade Bay, η is the prescribed tidal water level at the open boundary to the North Sea.

For other systems, the influence of the initial condition is dominant and the solution is given by

$$\psi(t) = \mathcal{F}[\psi(t_0), t - t_0] \tag{2.10}$$

for $t > t_0$. For chaotic systems this solution is only useful for times smaller than the predictability time τ_*. In both situations, the equations (2.10) and (2.9) forecast the future state of the system. These are *forecasts of the first kind*.

The most prominent example of this type of forecast are weather forecasts. Note that the progress made by the National Meteorological Center in predicting weather for several days in advance (see Sect. 5.1.3 and Fig. 5.6) comes not only from improving the numerical model $\mathcal{F}$, but equally from an improvement of the determination of the initial state $\psi(t_0)$.

2.6.3 Forecast of the Second Kind

Predictability arises from both the knowledge of the initial state and the external forcing. After the limit of predictability τ_*, the initial state carries no more predictive value. For longer times the deterministic operator $\mathcal{F}$ may be replaced by a stochastic operator $\mathcal{R}$. The solution $\psi(t)$ then takes the form

$$\psi(t) = \mathcal{R} + \int_{-t_0}^{t} d\tau \; \mathcal{G}(\tau - t)\eta(\tau) \tag{2.11}$$

and is now a random variable as well. The stochastic process $\mathcal{R}$ does not only represent the effect of the not-exactly known initial conditions but may also represent other processes that have not been accounted for in detail, as discussed in Sects. 2.1.4 and 2.5.

Once a stochastic description has been adopted, the type of forecast changes. In general, the parameters of the stochastic process $\mathcal{R}$, such as the mean, standard deviation and lag-correlation, depend on η. Assume, however, for simplicity that $\mathcal{R}$ is independent of the external forcing and that it has zero mean. Then, the state variable $\psi(t)$ becomes a random variable, which is conditioned[15] by the external forcing. Any forecast becomes a forecast of the statistical distribution of ψ. In practical terms this is mostly the conditional expectation

$$E(\psi(t)|\eta(\tau)) = \int_{-t_0}^{t} d\tau \; \mathcal{G}(\tau - t)\eta(\tau) \tag{2.12}$$

This type of forecast is called *forecast of the second kind*. They specify conditional probability distributions. Since time-averaging reduces the role of the "noise" $\mathcal{R}$ in many not-too-nonlinear systems, the (conditional) expectation of an ensemble of forecasts has often a smaller error than the individual forecasts[16].

Seasonal climate forecasts are an example of such forecasts of the second kind. Aiming at forecasting the seasonal mean of temperature or precipitation, an atmospheric model first exploits the information contained in the initial state, and later, after 10 days or so, the information provided by the persistent tropical sea surface temperature. When run in an ensemble mode, such forecasts attempt to simulate the probability distribution explicitly. These distributions are conditioned only weakly by the initial state and strongly by

---

[15] A random variable is "conditioned" if its parameters, like the mean or the variance, depend deterministically on some external variables.

[16] If the real evolution is considered a random realization of the (correctly predicted uni-modal) ensemble of possible realizations, then an exact forecast is not possible. In that case, it is best to offer the mean of the ensemble as the best guess of the actual evolution. In fact, the expected mean square error between the real evolution and a randomly chosen forecast is twice the mean square error between the real evolution and the mean evolution.

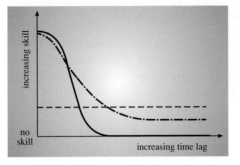

Fig. 2.3. Sketch of the predictive skill of systems that evolve under internal chaotic dynamics and depend on (inexactly known) initial states ((2.10), *continuous line*), of systems that are mainly controlled by external forcing ((2.9), *dashed line*), and of systems that are both under internal and external control ((2.11), *dashed-dotted line*)

the sea surface temperature. Sometimes other factors can be used for conditioning such forecasts, as for instance the presence of volcanic aerosols in the stratosphere.

The predictive skills for the three different cases, (2.9, 2.10, 2.11), are qualitatively sketched in Fig. 2.3. The skill of a system evolving mostly under internal chaotic dynamics drops off to "no skill" within its limits of predictability, while a system under the control of known external forces keeps an almost constant level of skill. A system controlled by both internal dynamics and external forces exhibits a high skill for initial times when initial conditions dominate and at later times a constant, usually low level of skill, when the external control is dominant. Of course, the relative importance of external and internal factors does not need to be stationary, so that the predictability may change with time. More quantitative measures of skill are introduced and discussed in Sect. 5.1.5 and Appendix C.2.4.

The determination of forecast skills and its dependence on exogenous conditions is one of the challenges of modern weather forecasting.

3

Models and Data

Models are one part of environmental sciences. Observational data are another part. Here we consider the interaction between these two parts. This interaction takes many forms. Here we consider

- the validation of models, where model results are compared with observational data to assess how "realistic" the model is (Sect. 3.1),
- data assimilation, where model results are combined with observational data to obtain a better field estimate or prediction (Sect. 3.2), and
- the calibration of models, where model parameter are estimated from data (Sect. 3.3).

3.1 Validation

Models are validated by model–data comparison. There are some "philosophical" problems with this comparison, which are discussed first. On a technical level model–data comparison is a statistical problem. Both the observational record and the model output are viewed as merely one realization of an ensemble of equally likely realizations consistent with the exogenous conditions. The comparison can take the form of a statistical test or the form of an estimate, estimating the "skill" of the model. A number of specific methodologies and concepts have been developed by the environmental science community.

A particular problem arises in model–data comparison and causes considerable confusion in the scientific community. What does the output of a model represent? In case of a grid-point model, does the number calculated at a grid point represent the real world value at that location? In our understanding, this is not the case. In view of our discussion of the Reynolds decomposition in Chap. 2 we suggest interpreting the value at a grid point as a mean value over the associated grid box or even over a somewhat wider neighborhood. In fact, the model equations, with all their parameterizations are not "the equations" of the system but the equations for the system at the chosen resolution (cf. Sect. 2.1.3). That is, the model does not simulate local values but mean values averaged over a certain neighborhood. The same rationale must be applied when spectral models or finite element models are used.

Observations, on the other hand, are mostly local observations; exceptions are observing systems such as satellites, which produce area-averaged data. Thus, most model–data comparisons suffer from a built-in inconsistency between the model data and the observation data. If the considered variable is sufficiently smooth, i.e., if it has a correlation length larger than a grid box diameter, then the point observation may be considered very similar to the grid box average. The correlation length may be very different for different variables. The geopotential height at 500 hPa in the atmosphere is a rather smooth field, it varies only over hundreds of kilometers, whereas convective precipitation changes strongly on spatial scales of a few kilometers. Even the same variable can have different correlation length in different circumstances. An example is sea level elevation, which is rather smooth in the open ocean, but near the coast line an area average can differ significantly from the measurement right at the shoreline (cf. Fig. 3.1). When the correlation length of the considered variable is short, then two approaches can be used: either an *observation model* is constructed to relate grid point values to local data (see Sect. 3.1.3), or a series of observations in the grid box is averaged.

3.1.1 Validation as a Philosophical Problem

Models, like theories and other cognitive constructs, can never be verified. Even if they correctly predict reality under one set of circumstances one cannot be sure that they will also do so under a different set of circumstances. The logician concludes from this fact that models can only be falsified but not verified. They are false if they do not "predict"[1] reality correctly. This categorical statements is, however, not appropriate for models of environmental systems where the notion of correct or incorrect prediction is not well-defined. The appropriate notion is how well does a model reproduce reality? The question is not whether a model is right or wrong but how good it is. Quantitative measures need to be developed for answering this question. The purpose of the model needs to be included. Verification also implies the concept that the model predicts reality for the right reasons. As discussed, we can never be sure about this because environmental systems are open [127], [132].

For the above reasons one introduces the weaker concept of validation. One only requires that the model results are consistent with observations. One does not claim that the model is "correct" but only that it "works".

---

[1] In many disciplines, the term "prediction" refers to attempts to describe future states. Here, however, the term is used in a more general sense, namely as an attempt to specify the outcome of an experiment that, e.g., simulates the frequency of certain extreme events.

A useful concept is that of analogs. Following the philosopher of science Hesse [61], models have positive, neutral and negative analogs with reality, or among each other. Positive analogs are common properties, and a validation strategy should show that they prevail both in reality and in the model. Neutral analogs are properties for which it is not known whether they are common properties, and negative analogs are properties that are not shared by model and reality. In case of climate models, positive analogs are the conservation of mass, energy and momentum, neutral analogs are the sensitivity to changing greenhouse gas concentrations, and negative analogs are the propagation of sound waves in the ocean or atmosphere and the existence of a time step in the numerical code. Because of the negative analogies, all models are "wrong" in a trivial sense. The task of validation is to determine the positive and negative analogs and to assess whether the extent of the positive analogs makes the model suitable for certain applications.

The added value of modeling comes from assuming that the neutral analogs are actually positive ones: that a response of a model to a forcing is actually the response that the environmental system would show if subjected to the forcing without any other changes. A forecast prepared with a model is hoped to coincide with the actual development to be observed in the future.

One aspect should be stressed. Even if a model is validated, i.e., the existence of a series of relevant positive analogs confirmed, there is no certainty that these analogs are still positive, when the model operates with parameters outside the range covered by the empirical evidence used for validating the model. If a climate model describes the present climate well, this is no *proof* that it describes paleoclimatic states or Global Warming well. There may be numerous good reasons to *believe* in a model's skill in doing so, but there remains always the possibility, albeit sometimes a small one, that relevant aspects of the non-observed part of the parameter space are not sufficiently taken into account.

3.1.2 Some Common Approaches

Different validation approaches are commonly in use:

A *model* is used to reproduce a certain time period with good observational coverage, often a special observation campaign. Then, the observations are directly compared with the numbers generated by the computer model. When such episodes are simulated satisfactorily, it is hoped that the model will do well also under other, non-observed conditions.

The model is checked whether it satisfies *cognitive models* that are known to be valid for the real world. Examples of such cognitive models are the conservation laws for mass, energy, momentum, enstrophy and angular momentum and the geostrophy of large scale flows. An example is Ulbrich's and Ponater's [167] comparison of the atmospheric energy cycle according to Saltzmann (see [130]) in a model simulation with analyses of the real circulation.

Forecast skills. As has already been pointed out, we restrict the word "forecast" to a model application where a dynamical model is used to predict in detail a future state, exploiting knowledge about the dynamics of the system and knowledge about the initial state (see Sect. 2.6). Examples are the standard weather forecast (Sect. 5.1.3), forecasts of storm surge water levels (Sect. 5.1.2) or of river streamflow, or of algae blooms in marine environments. In Sect. 5.1 we discuss different forecasts and their interpretations.

All these forecasts are not perfect but suffer regularly from mispredictions. Forecasts generally have only a limited forecast skill. As discussed in Sect. 2.6 this limitation may be due to chaotic dynamics, as in the case of the weather or algae bloom forecast, or due to the inability to correctly prescribe atmospheric drivers like wind or precipitation, as in the case of river streamflow or storm surges.

The forecast is a number (or an array of numbers). To define skill measures, it needs to be compared with a corresponding observation. To do so, the two variables are considered realizations of a bivariate random variable. When the forecast is done repeatedly for different situations then one has many realizations or samples {(forecast 1,observation 1), ..., (forecast n,observation n)}, on which statistical analyses can be performed. The model is considered skillful, if the two components, observed and predicted, of the bivariate random variable co-vary closely. Erratic, unrelated variations are indicative that the forecast is not useful, either because the model is insufficient or because the forecast time is beyond the predictability limit. A number of such statistical skill measures are in use. Some of them are discussed in Sect. 5.1.5 and Appendix C.2.4.

In *simulations*, there is no direct correspondence between the time in reality and the time in the model. Instead simulations generate possible trajectories in phase space, which are considered possible outcomes of the real world. Consistency with the record of real world observations is assessed by comparing statistics generated by the model with the same statistics derived from the observational record. Such statistics can be simple means and variances, but can also be more complex quantities, like covariances between state variables or between a state variable and a forcing factor, time lag covariances, spectra and empirical orthogonal functions (see Sect. C.2).

Of course, it is best that a model be validated with data representative for as large a part of the parameter space as possible. Thus, tide models should be run for many different bathymetric and coastline configurations, and climate models should be used for seasonal forecasting and paleoclimatic reconstructions.

3.1.3 Validation as a Statistical Problem

Mathematically, validation is a statistical problem. No model prediction ever coincides exactly with observations. To make a meaningful comparison one has to assign errors both to the model and the observations.

Formally, one combines the simulated state vector ψ_i for all time steps $i = 1, \ldots, K$ into the model vector $\psi = (\psi_1, \ldots, \psi_K)$. The dimension of the model vector is the dimension of the state vector times the number of time steps. This model vector is regarded as an estimate of the true model vector ψ_t with an associated error γ. Thus

$$\psi = \psi_t + \gamma$$

The error γ comprises all the errors introduced into the model by approximate equations, parameterizations, inexact boundary and forcing fields, numerical discretization and other uncertainties.

Similarly, we combine all observations into an observation vector ω. Its dimension is the number of observations. The observation vector is regarded as an estimate of the true vector ω_t with an associated error λ. Thus

$$\omega = \omega_t + \lambda$$

The error λ accounts for instrumental, environmental and sampling errors. Most often the state vector of the dynamical model cannot be measured in its entirety. Measurements are usually available only at a few locations and at a few times, and these locations and times usually do not coincide with the grid points and time steps of the dynamical model. Furthermore, model variables at grid points and time steps must be interpreted as averages, as we pointed out. In addition, observations often do not consist of measurements of state variables but of some other quantity which is related to the state variables. The state variable in a quasi-geostrophic model is the streamfunction which must be related to temperature and salinity if only these are observed. Even more extreme are proxy data such as lake warves or tree rings which must be related to water level and temperature. All these circumstances are accounted for by introducing an observation equation

$$\omega_t = \mathcal{C}[\psi_t] \tag{3.1}$$

where the operator $\mathcal{C}$ relates observation and model. In many cases $\mathcal{C}$ is just a projection, i.e., an operator that selects from the values at all grid boxes the values for those grid boxes where measurements are taken. However, when the local observations do not coincide with a grid box value, or when the local observation do not represent a state variable at all, then $\mathcal{C}$ becomes a more complicated operator and specific efforts are needed to specify it. An example of an observation model is shown in Fig. 3.1 where the observed water level ω at a coastal tide gauge is related to the near coastal sea level ψ calculated in a model. According to this observation model, the water level measured at

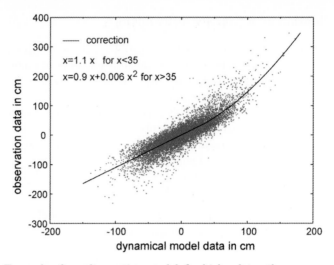

Fig. 3.1. Example of an observation model $\mathcal{C}$ which relates the near coastal water level ψ calculated from a North Sea model (*horizontal axis*) to the observed shoreline water level ω (*vertical axis*). Reprinted from [95] (© (1999), with permission from Elsevier)

the shore is somewhat higher than in the open water for moderate sea levels; for storm surge levels the water level at the coast line rises significantly above the level calculated in the open water.

Inserting the model vector ψ into the observation equation (3.1) gives a model estimate for the observations

$$\omega^m = \mathcal{C}[\psi]$$

This model estimate needs to be compared with the actual observations ω. This is a statistical problem. Different types of statistics can be considered, such as moments, covariances, teleconnection patterns and different types of comparisons can be performed, such as confidence bands, hypothesis testing and recurrence analysis. Some of these statistics and comparisons are discussed in Appendix C. Figure 3.2 sketches a particular comparison, the comparison of means. Another particular statistics is the skill of a forecast which is a quantitative measure to assess the usefulness and relative merits of different forecast schemes. Applications are given in Sect. 5.1.3 and technical details in Appendix C.2.4. The above formalism can easily be generalized to account for the fact that the operator $\mathcal{C}$ relating model and observations is not exact but also contains errors.

VALIDATION

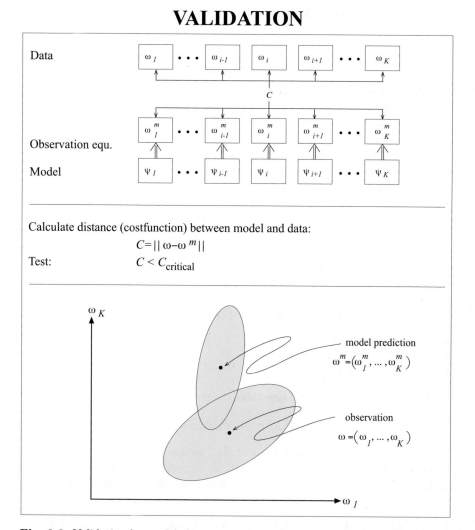

Fig. 3.2. Validation by model–data comparison. Given a set of observations $\boldsymbol{\omega}$ and model predictions $\boldsymbol{\psi}$ one first uses the observation equation to calculate the model estimate $\boldsymbol{\omega}^m$ of the observations and then considers the distance $C = \parallel \boldsymbol{\omega} - \boldsymbol{\omega}^m \parallel$ between $\boldsymbol{\omega}$ and $\boldsymbol{\omega}^m$. If this distance C is smaller than a critical value $C_{critical}$ then the model and data are deemed to be consistent. Distances and critical values are often based on the covariance matrices of $\boldsymbol{\omega}$ and $\boldsymbol{\omega}^m$, given by the ellipsoids. Model and data are then consistent when the ellipsoids sufficiently overlap

3.2 Data Assimilation

Data assimilation (DA) refers to techniques that combine dynamical models with observational data in order to obtain a better product. The better product might be a better estimate of a field, a better estimate of a model parameter, a better prediction, or a better substitute reality. Observationalists use dynamical models to extract reliable information out of observational data. DA then becomes "dynamically consistent interpolation". Modelers use observations to force their models to stay close to reality and control the loss of predictability. DA then becomes the *data analysis* of operational weather forecasting. These and other DA techniques have a common conceptual and mathematical background which comes from estimation and control theory. Here we describe the basic concepts of DA. Details are given in Appendix D. Examples of DA are given throughout the book.

Philosophically, DA is the farthest away from the classical paradigm that separates measurements and observations from physical laws and theory. It not only realizes that models require data to yield useful predictions but also that data require models to extract useful information. The utility of data becomes model dependent.

Data assimilation (DA) consists of three basic components:

- a dynamical model,
- a set of observations, and
- an assimilation scheme.

For the dynamical model one takes

$$\psi_{i+1} = \mathcal{A}_i[\psi_i; \boldsymbol{\alpha}] + \boldsymbol{\epsilon}_i \tag{3.2}$$

where i is the discrete time index and $\mathcal{A}_i$ a generally nonlinear operator that comprises the dynamics of the system. It depends on a set of parameters $\boldsymbol{\alpha} = (\alpha_1, \ldots, \alpha_L)$. The model (3.2) differs from (2.4) by the error term $\boldsymbol{\epsilon}_i$ which takes into account effects like (cf. Sects. 2.1 and 2.2):

- the operator $\mathcal{A}_i$ is inexact because it does not include all dynamical processes,
- the boundary conditions and external forcing fields are not exactly known,
- the spatial discretization requires uncertain parameterization of subgrid scale processes, and
- the discretization in time introduces truncation errors.

The second component of DA is a set of observations. If the observations are combined into an observation vector $\boldsymbol{\omega}_i$ then DA assumes that there exists an observation equation

$$\boldsymbol{\omega}_i = \mathcal{C}_i[\psi_i] + \boldsymbol{\delta}_i \tag{3.3}$$

which relates the observation vector $\boldsymbol{\omega}_i$ to the state vector ψ_i. Here $\mathcal{C}_i$ is again a generally nonlinear operator and $\boldsymbol{\delta}_i$ the observational error, representing instrumental, environmental and sampling errors and other uncertainties. The time index i must be identical in the dynamical model 3.2 and the observation model 3.3. In most cases, the time interval into which observations have been binned is much longer than the time step of the numerical algorithm used to solve the dynamical model. In this case, the time step of DA is thus given by the sampling procedure of the data and not by the original time step of the dynamical model. The collection of data into time bins also contributes to the error δ_i. The dimension of the state vector is M. The dimension of the observation vector is N. In environmental sciences one is generally in a situation where one has sparse data, $N \ll M$. The observations represent a weak constraint on the model. An example of an observation model was shown in Fig. 3.1.

In the case $N < M$ the operator C_i cannot be inverted to yield the state vector ψ_i. The state vector is underdetermined. Additional constraints are needed to arrive at a unique solution. In DA the additional constraint is consistency with a dynamical model.

The third component of DA is a blending scheme where one combines the model prediction ψ_i with the observation ω_i to form an improved estimate $\hat{\psi}_i$. The blending scheme depends on whether one wishes to solve a filtering problem or a smoothing problem. Assume that data and model predictions are available for times $i = 1, \ldots, K$.

A filtering problem combines past and present data to form a "nowcast". Formally, the nowcast $\hat{\psi}_i$ at time i is derived from all ψ_j and ω_j with $j \leq i$. Such a nowcast can be used to provide optimal initial conditions for a forecast. Later, when data and forecast have become available for the time $i + 1$, the procedure can be repeated to estimate $\hat{\psi}_{i+1}$, and so forth.

A smoothing problem combines all data, past, present and future, to form an optimal estimate for all times. Formally, $\hat{\psi}_i$ for all $i = 0, \ldots, K$ is derived from all ψ_j and ω_j with $j = 0, \ldots, K$.

Although both problems are estimation problems the filtering problem has its methodological roots in estimation theory and the smoothing problem has its roots in control theory.

Given the statistics of model error ϵ_i and the observation errors δ_i one can for any given blending scheme calculate the statistics of the error of the estimate $\hat{\psi}_i$. Optimal blending schemes can then be determined that minimize this error. This is not as straightforward as it sounds:

- Assigning quantitative error statistics to models and data involves considerable subjective judgements. What error is introduced into a dynamical model because it only represents approximate dynamics and parameterizes subgrid scale processes? What errors are introduced into an observational model because observations are not taken at grid points or represent proxy data?

- If nonlinear blending schemes are allowed for, then it is not sufficient to characterize the model and observation errors by just their means and covariances. Higher-order moments or the full probability distributions are needed.

- Any minimization requires the specification of a distance or norm, often in the form of cost, penalty or risk functions. As these names indicate, there is no unique objective way to arrive at such a distance. What is costly for one purpose might not be costly for another purpose.
- The minimization problem might not have a (unique) solution or the solution cannot be determined.
- Even if an optimal blending scheme can be determined it might be impractical, computationally excessive or suffer some other drawback.

For the above reasons a variety of adhoc and heuristic DA schemes have been developed. Most of them can be put into a common framework which is outlined in Appendix D. This framework is sketched in Figs. 3.3 and 3.4. When using any of these schemes it is paramount to

- understand the a priori statistical assumptions that go into the scheme, and
- determine the error of the blended estimate.

Methodologically, DA lies between validation and quality control. In validation the model prediction and the observed data are compared without any feedback. In DA, model prediction and observational data are combined to form an improved estimate. In the filtering problem, data are fed into the model to improve the forecast. In the smoothing problem, the model is fed into the data to extract dynamically consistent information. In both problems the data are given. Control theory also determines which data need to be taken for optimal control. In our context this would be the determination of an optimal observational sampling strategy.

3.3 Calibration

Dynamical models

$$\psi_{i+1} = \mathcal{A}_i[\psi_i; \boldsymbol{\alpha}] \tag{3.4}$$

generally contain a set of parameters $\boldsymbol{\alpha}$ that characterize the system and its interaction with its surroundings. These parameters are coefficients and boundary conditions in the original dynamical equations. In an ocean general circulation model they include the thermodynamic coefficients of sea water (such as the specific heat), the fluxes of momentum, heat and fresh water across the air–sea interface, the eddy diffusion coefficients that arise from Reynolds averaging, the topography, the earth's rotation rate and gravitational acceleration, and the tidal potential.

FILTERING

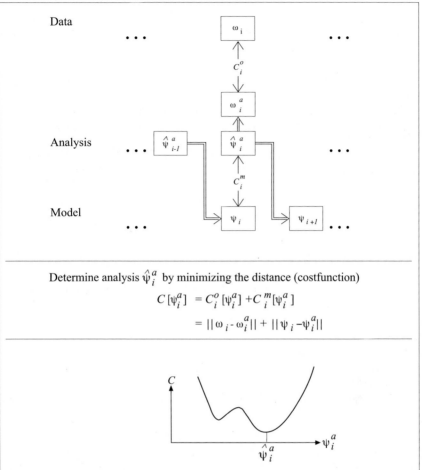

Determine analysis $\hat{\psi}_i^a$ by minimizing the distance (costfunction)

$$C\,[\psi_i^a] \;=\; C_i^o\,[\psi_i^a] + C_i^{\,m}\,[\psi_i^a]$$

$$=\; \|\,\omega_i - \omega_i^a\,\| + \|\,\psi_i - \psi_i^a\,\|$$

Fig. 3.3. Sequential data assimilation by filtering. At each time step i the analysis $\hat{\psi}_i^a$ is obtained from the data $\boldsymbol{\omega}_i$ and the model values $\boldsymbol{\psi}_i$ by simultaneously minimizing the weighted distance $C_i^o = \|\,\boldsymbol{\omega}_i - \boldsymbol{\omega}_i^a\,\|$ between data and analysis and the distance $C_i^m = \|\,\boldsymbol{\psi}_i^a - \boldsymbol{\psi}_i\,\|$ between analysis and model. The analysis $\boldsymbol{\psi}_i^a$ is used as the initial condition for the calculation of the model value $\boldsymbol{\psi}_{i+1}$ at time step $i+1$

SMOOTHING

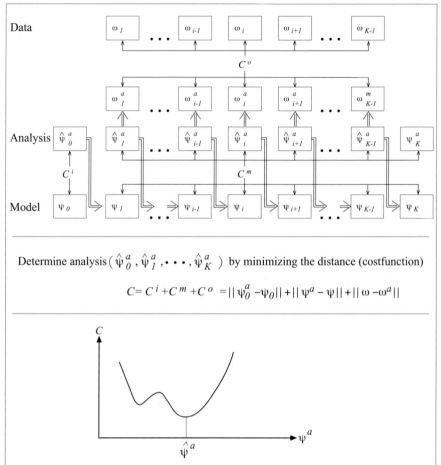

Determine analysis $(\hat{\psi}_0^a, \hat{\psi}_1^a, \cdots, \hat{\psi}_K^a)$ by minimizing the distance (costfunction)

$$C = C^i + C^m + C^o = \| \psi_0^a - \psi_0 \| + \| \psi^a - \psi \| + \| \omega - \omega^a \|$$

Fig. 3.4. Data assimilation by smoothing. An optimal field estimate or analysis $(\hat{\psi}_0^a, \hat{\psi}_1^a, \ldots, \hat{\psi}_{K-1}^a)$ is obtained by simultaneously minimizing the distance $C^i = \| \psi_0^a - \psi_0 \|$ between analysis and guess of initial condition, the distance $C^m = \| \psi^a - \psi \|$ between analysis and model prediction, and the distance $C^o = \| \omega - \omega^a \|$ between data and analysis. The adjoint method assumes $C^m = 0$. In this case the analysis ψ_i^a is a solution of the dynamical model for $i = 1, \ldots, K$ and the initial condition ψ_0^a is the only control variable

In order to execute the model these parameters need to be specified by the user. This includes sensitivity studies where the sensitivity of the model output to changes in parameters is investigated. In this case the model will be executed for different but specified values of the parameter. To asses the effect of the heat flux on a certain circulation pattern one would execute the model for various different prescriptions of the heat flux.

Though the parameters need to be specified it does not imply that they are well known. Some of them are, such as the earth's rotation rate and gravitational acceleration; others are not, especially those parameters that come about by closing the system, such as eddy diffusion coefficients. These not so well known parameters are often calibrated or tuned by comparing model results with observational data. This is again a statistical estimation procedure, similar to DA but with the parameter α now being a control parameter [2]. An optimal value of the parameter α is determined by minimizing the distance between the observations ω and the model output $\omega^m(\alpha)$. In practice, this is often done through trial-and-error, performing calculations for a limited number of parameters and parameter values. There are also examples when a rigorous minimization is performed (e.g., [146]), but never for the complete suite of all relevant parameters. Technical details are given in Appendix D.3.3.

Though such calibration or tuning of models is widely used three cautionary notes are in order:

- The data that have been used for calibrating a model cannot be used for validating a model. Calibration and validation has to be done with independent data.

- A model calibrated by a set of data can be expected to perform optimally under circumstances represented by this set of data but not necessarily for other circumstances. Diffusion coefficients obtained by calibrating an ocean circulation model to observation in the North Atlantic over the past ten years might not be optimal for the North Pacific or the future ten years. Of course, it is just this very step, in going from a situation where the model has been calibrated (and validated) to a situation where it is not, that generates new knowledge about the system, but this step involves inherent risks, as we have stressed all along.

- Calibration of, say eddy diffusion coefficients, usually produces numbers for these coefficients whereas we expect them to be functions (or formulae) that express their dependence on the flow field. These functions are not obtained by calibration. One can, of course, assume a certain functional form for the eddy coefficients with a certain number of free parameters and then calibrate these parameters, but the functional form has to be specified a priori by the user. Calibration does not elucidate the underlying physics, but it is the understanding of this underlying physics that gives

---

[2] A control parameter or variable is a variable that is varied in order to find the minimum of the costfunction.

us the confidence to apply the model outside validated parameter ranges. Any model that aspires to predictive capabilities must base its parameter specifications on an understanding of the underlying physics rather than on mere tuning.

In summary: Calibration encodes observations into models and results in improved models; validation compares observations and models and improves our confidence in models; data assimilation combines observation and models and results in better forecasts and field estimates.

4

The Dynamics of Tides and Climate

We refer in our discussion of computer models in environmental sciences mainly to two examples, which we consider illustrative and useful for demonstrating our line of reasoning. The first case is the tides. They represent within limits a clear-cut, well-understood almost classical physical system. Its understanding has resulted in numerous important societal applications. The second case is climate. It is considerably more complex. It comprises not only physics but also various other natural sciences such as geology or ecology. It is under the influence of significant nonlinearities and stochasticity and can hardly be considered a classical system. Its functioning is not yet fully understood. Results from studies of climate have excited various societal responses. Different from the tides, these results are not only of technical nature but are often loaded with controversial economic and behavioral implications.

In Sect. 1.2 we discussed features that make environmental systems different from classical systems such as electric circuits or the thermodynamics of gases. These features are the presence of an infinite number of processes with non-uniform properties and interactions, which operate on many different spatial and temporal scales. In climate modeling, many of these processes are highly significant. In tidal modeling only a few are significant. This fact is sketched in Fig. 4.1. We have also added the number of significant processes needed to model marine ecosystems. These models require the inclusion of even more processes than tidal and climate models. Furthermore, there is no consensus which processes must be considered key processes and which variables must be considered state variables. The relatively poor performance of models of marine ecosystems is thus not surprising.

In Sects. 1.3 and 1.4 we have given a general overview about tides and climate: the basic phenomena, the history of ideas, and various modeling approaches. In the following two Sects. 4.1 and 4.2 we sketch in more detail what we know about the *dynamics* of tidal and climate systems[1].

---

[1] The text on the tides has been supplied to large extent by Jürgen Sündermann, Hamburg.

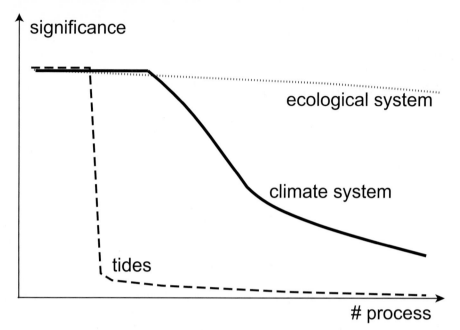

Fig. 4.1. Number of significant processes in different environmental systems

4.1 The Tidal System

4.1.1 The Nature of Tides

Tides are forced oscillations of the water masses of the world ocean under the influence of the gravitational and centrifugal forces in the celestial three-body system earth–moon–sun. The restoring force is earth's gravity. The specific oscillation patterns of the world ocean are controlled mainly by its topography and the earth's rotation (Coriolis force), much less by viscosity and bottom friction. The periods of tides are completely, and very accurately, determined from their celestial origin. If one develops the tide-generating potential into spherical harmonics, the periods of the dominating modes group into bands: semi-diurnal, diurnal, fortnightly, semi-annual (see Fig. 1.7). The amplitude of the corresponding partial tide in a specific region of the world ocean depends on the resonance behavior of free gravitational waves on the rotating earth (mainly Kelvin waves) in that region. Normally, the semi-diurnal waves dominate (as they should on the basis of celestial mechanics), but there are regions with mainly diurnal tides or even nearly without tides. In the case of local resonance the tidal ranges and velocities can become extremely high (see Fig. 1.1) and tidal power plants may be profitable.

The tide-generating forces are volume forces, which are acting on each particle in the water column. Correspondingly, the whole water column is affected, and the tidal currents vary little from the surface to the bottom, except for a shallow, bottom frictional layer (Fig. 4.2). Dynamically, the movement of water masses by the forcing fields of the moon and sun causes an inclination of the sea surface, representing a barotropic pressure gradient. The tides belong to the class of long gravity waves (water depth ≪ wave length) for which the hydrostatic approximation holds. The concepts of barotropy and hydrostacy imply that the pressure gradient is constant within the entire water column and, consequently, the velocities are constant. This vertical homogeneity renders the tides an essentially two-dimensional horizontal phenomenon. Correspondingly simple two-dimensional tidal models can be designed, without compromising realism.

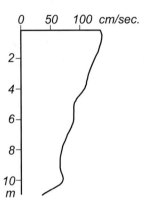

Fig. 4.2. Vertical profile of tidal velocities. After Sverdrup [164]

For the generation of tides by the celestial forces a sufficiently large water mass is needed. Such *autonomous* tides are generated only in the three oceans. Shelf seas with wide open entrances are affected by tidal waves penetrating from the deep ocean. These tides are called *co-oscillating* tides. Smaller seas with little or no connection to the world ocean (like the Mediterranean, the Baltic or the Caspian Sea) have almost no tides. Therefore, the tide generating forces must be implemented in a global tidal model as a forcing term in the momentum equation, making this equation a nonhomogeneous differential equation. In a shelf sea model the direct celestial forcing can be neglected, but the incoming tidal wave must be specified at the boundaries of the model area. The model thus takes the mathematical form of a set of homogeneous differential equations with a transient boundary condition at the open entrance.

4.1.2 Laplace Tidal Equations

Quasi-realistic modeling of the tides is relatively simple. Modeling other aspects of the physics of the ocean, like the wind driven circulation, is more complex. For tides one only needs:

- Laplace tidal equations,
- the topography of the ocean basin,
- the tide generating potential, and
- boundary conditions.

Laplace tidal equations are based on the shallow water equations with two additional major assumptions. First, they assume that the density of sea water is constant. Second, they neglect all oceanic motion other than of tidal origin. Since the gravitational force is a volume force and approximately constant throughout the water column, the tidal currents are depth independent. It hence suffices to consider the vertically averaged or barotropic horizontal velocity

$$\hat{\boldsymbol{u}}_h = \frac{1}{h + \xi} \int_{-h}^{\xi} dz \, \boldsymbol{u}_h \tag{4.1}$$

where ξ is the surface displacement, u_h the horizontal velocity and h the depth of the ocean. The equations for $\hat{\boldsymbol{u}}_h$ and ξ are obtained from the depth-averaged horizontal momentum and volume (or mass) balance. They are Laplace tidal equations:

$$\frac{D}{Dt}\hat{\boldsymbol{u}}_h + f\hat{\boldsymbol{z}} \times \hat{\boldsymbol{u}}_h + R\hat{\boldsymbol{u}}_h - A_h \Delta_h \hat{\boldsymbol{u}}_h = -g\boldsymbol{\nabla}_h \xi - \boldsymbol{\nabla}\phi_T$$

$$\partial_t \xi + \boldsymbol{\nabla}_h \cdot [(h + \xi)\hat{\boldsymbol{u}}_h] = 0 \tag{4.2}$$

with boundary conditions

$$\hat{\boldsymbol{u}}_h \cdot \boldsymbol{n} = 0 \;\; \text{at coastlines} \tag{4.3}$$

$$\xi = \xi_0(t) \;\; \text{at open boundaries} \tag{4.4}$$

Here $\frac{D}{Dt} = \partial_t + \hat{\boldsymbol{u}}_h \cdot \boldsymbol{\nabla}_h$ is the material or advective derivative, f the Coriolis parameter, $\hat{\boldsymbol{z}}$ the vertical unit vector, R a bottom friction coefficient, A_h a horizontal eddy viscosity coefficient, g the gravitational acceleration, ϕ_T the tidal potential of the moon and sun, $\boldsymbol{n}$ the normal vector of the coastline and $\xi_0(t)$ a prescribed water level. The bottom friction coefficient is either given by the nonlinear form

$$R = \frac{c_D}{(h + \xi)}|\hat{\boldsymbol{u}}_h| \tag{4.5}$$

in the spirit of the drag law (Appendix A.8.1) or assumed constant (linear bottom friction). Often the tidal potential is written as

$$\phi_T = g\xi_{equ} \tag{4.6}$$

where ξ_{equ} is the equilibrium tidal displacement.

In order to solve Laplace tidal equations one has to specify the parameters f, R or c_D, A_h and g and prescribe the tidal potential ϕ_T or equilibrium tidal displacement ξ_{equ}, and the surface elevation $\xi_0(t)$ at open boundaries. Laplace tidal equations contain nonlinearities in the advection and bottom friction terms and in the volume balance. For reasonable parameter ranges the nonlinearities are, however, weak and do not cause any chaotic behavior. Solutions of Laplace tidal equations are thus predictable. They are highly complicated, however, as a consequence of complicated bottom topography and coastlines.

If one introduces the depth integrated volume transport

$$\boldsymbol{U}_h = \int_{-h}^{\xi} dz \, \boldsymbol{u}_h \tag{4.7}$$

instead of the vertically averaged velocity $\hat{\boldsymbol{u}}_h$, the volume balance becomes strictly linear

$$\partial_t \xi + \nabla \cdot \boldsymbol{U}_h = 0 \tag{4.8}$$

The momentum equation is then usually approximated by

$$\partial_t \boldsymbol{U}_h = -f\hat{\boldsymbol{z}} \times \boldsymbol{U}_h - gH_0 \nabla_h(\xi - \xi_{equ}) - \boldsymbol{F}_h \tag{4.9}$$

where H_0 is the undisturbed water depth and $\boldsymbol{F}_h$ represents the effect of all frictional forces (bottom and lateral friction). All nonlinearities are eliminated, except for a possibly nonlinear bottom friction term. This change of representation and approximation often has considerable physical and numerical advantages (see Sects. 6.3.3 and 5.2.4).

In a numerical model Laplace tidal equations are spatially discretized and integrated forward in time until initial transients have died away. The state vector ψ thus consists of the elevations $\xi_i(t)$ and volume transports $\boldsymbol{U}_{h,i}(t)$ at all grid points $i = 1, \ldots I$. The dimension of the state vector is hence three times I. Often, forcing at a single tidal frequency is considered.

When a strictly linear version of the tidal equations is assumed then one can solve the equations by expanding the tidal potential $\xi_{equ}(t)$, the tidal elevations $\xi_i(t)$ and the tidal transports $\boldsymbol{U}_{h,i}(t)$ into a sum of tidal harmonics $exp(-i\omega_j t), j = 1, \ldots, J$. The complex tidal amplitudes $\xi_i(\omega_j)$ and $\boldsymbol{U}_{h,i}(\omega_j)$ can then be calculated for each tidal component j separately. The time derivative $\partial/\partial t$ is replaced by $-i\omega_j$ and the problem becomes purely algebraic. No time stepping is required.

4.1.3 Tidal Loading and Self-attraction

Because of the relative simplicity of the equations and the absence of strong nonlinearities, the first numerical tidal models have already provided convincing results [53], [9], [131], [193], [148]. Later the high accuracy demands of regional water level forecasts for navigation and coastal protection and the needs for correcting geodetical data demonstrated some deficiencies in global tidal models. A closer inspection of the problem led to the finding that the conventional Laplace tidal equations needed to be modified to account for three more processes, namely the *earth tides, tidal loading* and ocean *self-attraction*. The tidal potential causes tides of the solid but elastic earth, called the earth tides, which deform the ocean bottom. The moving tidal water bulge also causes an elastic deformation of the ocean bottom, the load tides. At the same time the water mass exerts variable gravitational self-attraction, depending on its own distribution. A numerical experiment with a global tidal model [193] confirms the hypothesis that these effects influence the global tides significantly. Figure 4.3 demonstrates this fact by showing tidal phases from simulations with and without these effects. Obviously the inclusion of earth tides, tidal loading and self-attraction brings the simulation much closer to the observed values, so that the conclusion is warranted that these three processes constitute first-order processes for the global tidal simulation problem. In shallow seas these effect are of minor importance. Technically, these effects can be included by modifying Laplace tidal equations as follows.

Denote the bottom elevation of the earth tide by ξ_e and the bottom elevation of the load tide by ξ_l. If these bottom elevations are included then the displacement in the pressure term of Laplace tidal equations is not the elevation ξ of the water column but the geocentric elevation

$$\xi_g = \xi + \xi_e + \xi_l \tag{4.10}$$

In addition, the earth tides, the load tides and the ocean tides redistribute mass and modify the gravitational potential (4.6) by amounts $\delta\xi^e_{equ}$, $\delta\xi^l_{equ}$ and $\delta\xi^o_{equ}$. This modification is the "gravitational self-attraction". The full tidal forcing is thus given by

$$\phi_T = g(\xi_{equ} + \delta\xi^e_{equ} + \delta\xi^l_{equ} + \delta\xi^o_{equ}) \tag{4.11}$$

The calculation of the earth and load tides must take into account the elastic properties of the solid earth. The calculation of the gravitational self-attraction involves a spatial convolution with a known Green's function. Under simplifying assumptions these calculations result in factors α and β modifying the elevations and equilibrium tidal displacement in the momentum balance (4.9)

$$\partial_t \mathbf{U}_h = -\mathbf{f} \times \mathbf{U}_h - gH_0\nabla_h(\beta\xi - \alpha\xi_{equ}) - \mathbf{F}_h \tag{4.12}$$

The factor α is usually taken to be 0.7 and the factor β to be 0.953 for semidiurnal and 0.940 for diurnal components[2]. Equations (4.12) and (4.8) represent the zeroth order physics: volume conservation, momentum changes

---

[2] The factor α arises from the earth tide ξ_e. It can be derived rather straightforwardly. Since the eigenfrequencies of the solid earth are much higher than the tidal frequencies, the earth tide is nearly in equilibrium and both the bottom elevation $\xi_e = h_2\xi_{equ}$ and the additional tidal potential $\delta\xi^e_{equ} = k_2\xi_{equ}$ are proportional to the equilibrium tidal displacement, ξ_{equ}, with coefficients $h_2 \approx 0.6$ and $k_2 \approx 0.3$. The inclusion of the earth tide thus introduces in the ∇_h-term two terms proportional to ξ_{equ}, and $\alpha = 1 + k_2 - h_2 \approx 0.7$.

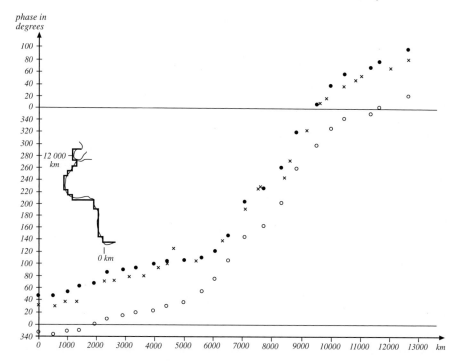

Fig. 4.3. Observed and computed phases of the M_2 tide along the west coast of Africa. (In degrees referred to the moon's transit at Greenwich). *Crosses:* observed values. *Filled/open circles:* computed values with/without tidal loading and ocean self-attraction [193]

due to the Coriolis, pressure, astronomical, and frictional forces. Solid earth tides, the load tides, and tidal self-attraction are included in an approximate manner. The frictional force consists usually of bottom friction, either linear or nonlinear, sometimes augmented by lateral friction.

4.1.4 The Tidal Inlet Problem

In Sect. 1.3.2 we introduced a hydraulic model (Fig. 1.4) of Jade Bay, a 10-km-wide tidal inlet in the Southern German Bight. A miniaturized, but still large model with a diameter of 10 m, was used to simulate the tidal water levels and currents. In the 1960s numerical models were constructed to replace these costly coastal engineering devices. In 1972 Sündermann and Vollmers made a

---

The factor β arises from the load tide ξ_l, which is caused by the weight of the overlying water column, and from the self-attraction terms $\delta\xi_{equ}^l$ and $\delta\xi_{equ}^o$ which all depend on the tidal elevation ξ. Equation (4.12) assumes all these terms to be just proportional to ξ and thus combine into a single factor β. A more accurate representation allows for different factors for different spectral harmonics.

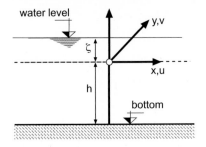

Fig. 4.4. Geometry and variables of the mathematical model of the tides in Jade Bay. From Sündermann and Vollmers [163]

systematic attempt to compare the hydraulic with a numerical model. Here we describe their model in some detail to demonstrate the actual structure of a tidal model.

Sündermann and Vollmers based their numerical model on Laplace tidal equations (4.2) with the boundary conditions (4.4). Figure 4.4 displays the Cartesian coordinates used: u is the velocity component in east–west direction x, v the velocity component in north–south-direction y, and ξ the water level relative to an undisturbed level h. They also made some simplifications. Since the water body of the Jade Bay is small, the tidal potential $\nabla \phi_T$ is neglected. Thus, the tidal system is forced through the boundary conditions only. Also neglected are the advective terms and the horizontal friction terms in the momentum balance. Bottom friction is nonlinear. In this coordinate system and under these assumptions Laplace tidal equations become

$$\partial_t u + \frac{c_D}{h + \xi} \sqrt{u^2 + v^2}\, u - fv + g\partial_x \xi = 0$$
$$\partial_t v + \frac{c_D}{h + \xi} \sqrt{u^2 + v^2}\, v + fu + g\partial_y \xi = 0 \qquad (4.13)$$
$$\partial_t \xi + \partial_x (h + \xi)u + \partial_y (h + \xi)v = 0$$

where the "hat" on the velocity has been omitted.

A major difference between the hydraulic system and the system given by Laplace tidal equations (4.13) is that the hydraulic system does not contain a scaled version of the Coriolis force. The tidal equations can be integrated with and without the Coriolis term. We come back to this issue in Chap. 6.

Next we present a few details of the discretization of the differential equations (4.13). First a suitable grid for the dependent variables, zonal and meridional currents u and v and water level ξ, must be chosen. Then the differential operator must be replaced by difference operators.

For the dependent variables the staggered C-grid is used (see Sect. B.3), as sketched in Fig. 4.5, with grid points marked by "+" for meridional currents (v), "×" for zonal currents (u) and dots "." for water levels (ξ).

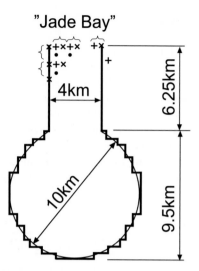

Fig. 4.5. C-grid used in the Jade Bay study, with $+$ = v-points, $\times$ = u-points and $\cdot$ = ξ-points. From Sündermann and Vollmers [163]

The discretization in time is forward and in space central[3]

$$\frac{du}{dt}\bigg|_n = \frac{u_{n+1} - u_n}{\Delta t}$$

$$\frac{du}{dx}\bigg|_n = \frac{u_{n+1} - u_{n-1}}{2\Delta x} \qquad (4.14)$$

$$\frac{du}{dy}\bigg|_n = \frac{u_{n+1} - u_{n-1}}{2\Delta y}$$

Stability is guaranteed if the Courant-Friedrich-Levy criterion

$$\Delta t \leq \frac{\Delta x}{\sqrt{2gh}}, \frac{\Delta y}{\sqrt{2gh}} \qquad (4.15)$$

is satisfied.

We come back to this case in Chap. 6.

---

[3] Historically, this scheme was named "HN"-method: hydrodynamic–numerical. Walter Hansen seems to have been the first to have introduced an HN approach in the 1930s and 40s. The first comprehensive documentation is [53].

4.2 The Climate System

4.2.1 Components and Processes

The main fluid components of the climate system are the oceans, the atmosphere and the cryosphere[4] (Fig. 4.6). They transport and exchange among each other matter – mainly water but also other gases and particles – and momentum and energy in such manner that the incoming solar radiation is, on average, balanced by the outgoing thermal radiation (see Fig. 1.11, and Sect. 4.2.4). Total mass and angular momentum are conserved[5]. Climate is a thermodynamic engine. Motions are driven by the contrast of cooling (net long-wave radiation loss) and heating (net short-wave radiative gain).

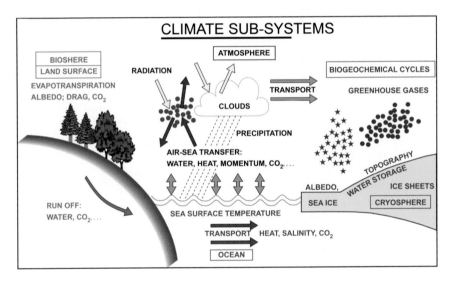

Fig. 4.6. Processes represented in a climate model. From von Storch and Hasselmann [178]

In the *atmosphere*, matter, momentum and energy are transported by the winds. Water vapor, clouds, other gases and aerosols modify the radiative properties of the atmosphere, by absorbing, reflecting and emitting radiative energy. Winds cause the formation of turbulent boundary layers above the land surface, the oceans and the ice sheets. Efficient fluxes of matter, energy and momentum couple the climatic components through these boundary

---

[4] Sea ice and even icesheets move and are hence fluids but have a very different rheology from air and water.

[5] If exchange of mass with space and of angular momentum with celestial bodies is disregarded.

layers. The atmosphere is mostly weakly stably stratified, but numerous situations exist where local vertical instabilities, often reinforced by condensation of lifting air, occur; therefore, the atmosphere as a whole is relatively well mixed.

The atmosphere represents relatively little mass and has little thermal inertia. Therefore, the atmospheric state at a given time is mostly in a statistical equilibrium with the thermal state at its lower boundary, i.e., at the sea surface, the land surface and the ice sheets. Nevertheless, the role of the atmosphere in the climate machinery is vital. The atmosphere controls the amount of energy absorbed from the sun and re-emitted into space. It provides mechanical energy to mix the ocean, momentum to set up the wind-driven currents, and creates density variations at the oceans' surface that drive the thermohaline circulation. Its high-frequency variability acts as stochastic forcing on the slower evolving components of the climate system.

The *ocean*, on the other hand, has much more mass and larger thermal inertia. Currents, with velocities much slower than those of the winds, transport momentum, energy, water, salt, and chemicals. The oceans store large amounts of matter, such as carbon dioxide. Near the surface a turbulent boundary layer is formed, which is in close contact with the overlying atmosphere and, to a lesser degree, with overlying sea ice. In the deep ocean, variations on time scales of days and weeks are fairly weak. The deep ocean has only limited contact to the upper ocean. The main process connecting the upper with the deep ocean is convection. It is triggered in sub-polar oceans when the surface waters become denser, as a result of cooling and accumulation of salt during sea ice formation. Tides are very relevant in marginal seas, where tidal currents reinforce the bottom boundary layers. Tides in the open ocean convert part of their energy into internal tides at topography. These internal tides are believed to provide a major part of the mechanical energy needed for diapycnal mixing[6] and maintenance of the abyssal circulation.

The *cryosphere* comprises ice sheets, like Antarctica or Greenland, ice shelves, sea ice and snow areas. Ice sheets appear to the human observer as constant, but they actually are fluids like the atmosphere and the ocean, but with much smaller velocities. Indeed, ice sheets are larger versions of glaciers. The size of glaciers is determined by their mass balance: at high altitudes, mass is accumulated through freezing precipitation; this mass is slowly moving downhill to lower altitudes, where the ice is melting. Whether the glacier as a whole accumulates or loses water, depends on its height distribution; above a certain height, precipitation freezes, and below it ice melts. An equilibrium is obtained, when the "accumulation" area times the precipitation rate equals the "ablation" area times the melting rate. Ice sheets evolve similarly. They efficiently store fresh water, controlling thereby not only sea level but also the concentration of salt in seawater. Ice sheets usually vary on time scales of thousands of years but paleoclimatic evidence indicates that

---

[6] Mixing across layers of constant density.

sometimes faster changes took place. Mountain glaciers exhibit marked variations on time scales of tens of years. Sea ice and snow fields have little inertia, and are mainly dependent upon the local atmospheric and oceanic conditions. They have a significant impact on the exchange of heat and matter between the atmosphere, ocean and land surface, and efficiently reflect sun light, thus affecting earth's energy balance.

There are more components to the climate systems. The land surface is one such additional component. It stores heat and, more importantly, water. The flux of energy and water is also affected by the presence of vegetation. Vegetation and soil – which may be seen as one dynamical system – affect biogeochemical cycles, since for instance carbon dioxide is stored by plants and released by litter. Thus, the biosphere is another significant part of climate.

People and society are also part of the climate engine, as human action determine modifications of the land surface and emissions of climatically relevant substances such as greenhouse gases and aerosols.

For variations on time scales of hundreds of thousands of years and longer, the lithosphere becomes a dominant factor. Weathering of rock, burial of sediments, formation of mountains, changing of land–sea distribution and other lithospheric processes have altered climate dramatically [20], [170].

From this list, it becomes obvious that climate is not a finite and closed system. An infinite number of processes, each with its own dynamics, are interacting. Modeling this system requires decision about which processes to retain and which to disregard. Some help comes from the fact that most processes are characterized by specific time and space scales, as discussed next.

4.2.2 Scales

In the previous subsection we have attempted an, albeit incomplete, overview of climate processes. These processes take place on all space and time scales; for the ocean and the atmosphere, this is sketched in Fig. 4.7. Molecular processes are disregarded. The shortest time scales are then associated with sound waves, micro turbulence and wind seas, which may have spatial scales as small as a few centimeters. Boundary layer processes have time scales of minutes, and spatial scales of meters. The largest scales of thousands of kilometers, have temporal scales of a few days in the atmosphere (planetary waves) and hundreds and thousands of years in the ocean (thermohaline or meridional overturning circulation). In both fluids small spatial scales are generally associated with fast time scales, and large scales with slow variations. This behavior is also reflected in the dispersion relations of many wave types (see Fig. A.4). It offers a natural ordering of processes. Processes in ice sheets can also be ordered in this way. But it should be stressed that this ordering is not a universal law for environmental systems. In particular, it does not apply to coastal systems, as can be seen in Fig. 4.8 which shows a much

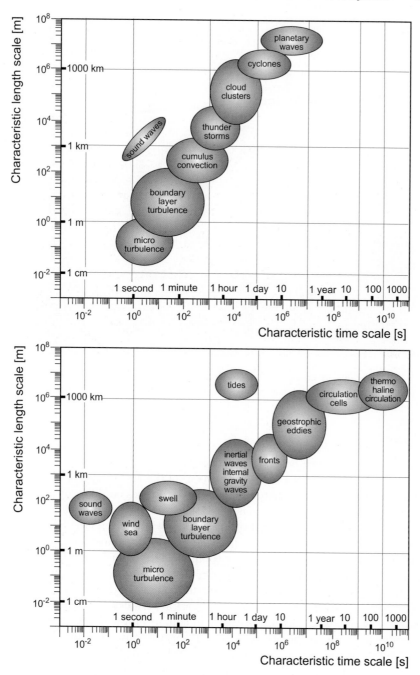

Fig. 4.7. Spatial and temporal scales of atmospheric and oceanic dynamics. From von Storch and Zwiers [182]

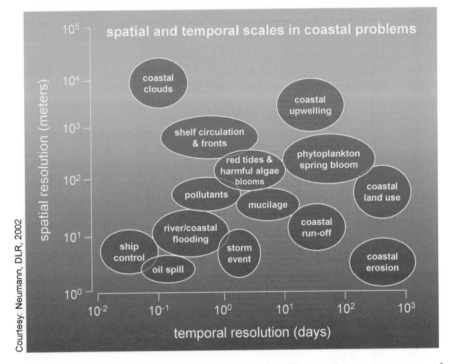

Courtesy: Neumann, DLR, 2002

Fig. 4.8. Scales in coastal research, demonstrating the simultaneous presence of all spatial and temporal scales in this complex environment. Courtesy: Nicolas Hoepffner

more complicated situation without any simple relation between spatial and temporal scales. Such ordering also does not apply to ecology.

An implication of the relation between spatial and temporal scales is that climate dynamics is dominated by different processes on different time scales. When short time scales are considered, such as daily and weekly variations of weather, then oceanic variations related to wind-driven circulation cells or the thermohaline circulation are irrelevant. Indeed, in weather forecasts only the atmosphere changes. The other components remain fixed. On time scales of years, the details of cloud clusters in the atmosphere, fronts in the ocean are as irrelevant as the thermohaline circulation or the dynamics of ice sheets. When time scales of many thousands of years are studied, then the atmosphere may be considered as being essentially in equilibrium with the slower climate components such as the ocean or the ice sheets, while lithospheric processes may be disregarded. On time scales of hundreds of thousands of years and longer, the dynamics of the oceans and of the ice sheets become also relatively unimportant compared to the geological processes modifying the surface of earth.

When modeling the climate system, in most cases an interval of spatial and temporal scales is considered. The larger scale dynamics is fixed. The smaller scale dynamics is parameterized.

4.2.3 The Primitive Equations

The differential equations used to describe the oceanic and atmospheric dynamics on large spatial scales are usually the so-called *primitive equations*. They are derived from the fundamental dynamical laws by a number of approximations. The most important one is the shallow water approximation. It consist of the hydrostatic approximation which neglects the acceleration in the vertical momentum balance and the traditional approximation which neglects the meridional component of the earth's rotation. The shallow water approximation is justified for motions whose aspect ratio, i.e., whose ratio of vertical to horizontal length scale, is much smaller than one. The state variables are the wind or current velocity vector, with the horizontal components $\mathbf{v}_h = (u, v)$ and the vertical component w, the density ρ, the specific humidity q (atmosphere) or the salinity S (ocean), temperature T, and pressure p. The equations are usually expressed in spherical coordinates with longitude θ, latitude φ and height z. For the atmosphere one such set of primitive equations is

$$\rho\left(\frac{Du}{Dt} - \frac{uv}{a}\tan\varphi - fv\right) = -\frac{1}{a\cos\varphi}\partial_\theta p + \mathcal{F}_u$$

$$\rho\left(\frac{Dv}{Dt} + \frac{u^2}{a}\tan\varphi + fu\right) = -\frac{1}{a}\partial_\varphi p + \mathcal{F}_v$$

$$0 = -\partial_z p - g\rho$$

$$\frac{D\rho}{Dt} + \rho(\nabla \cdot \mathbf{v}) = 0 \qquad (4.16)$$

$$\rho\frac{Dq}{Dt} = \mathcal{G}_q$$

$$\rho c_p\frac{DT}{Dt} = \mathcal{G}_T + \alpha\frac{Dp}{Dt}$$

$$p = R_m(q)\rho T$$

where

$$\frac{D}{Dt} = \frac{\partial}{\partial t} + \frac{u}{a\cos\varphi}\frac{\partial}{\partial\theta} + \frac{v}{a}\frac{\partial}{\partial\varphi} + w\frac{\partial}{\partial z}$$

is the material or advective derivative[7], f the Coriolis parameter, a the radius of earth, c_p the specific heat, α the thermal expansion coefficient, and R_m the gas constant. The terms $\mathcal{F}_u$, $\mathcal{F}_v$, $\mathcal{G}_q$ and $\mathcal{G}_T$ represent the parameterized effects of unresolved processes. The first two equations are the horizontal

---

[7] It is the material derivative only for scalar fields. For the velocity components $\frac{D}{Dt}$ plus the $\tan\varphi$ terms form the material derivative.

components of the momentum balance and the third equation is the vertical momentum balance, approximated by the hydrostatic balance. The next equation is the mass balance or continuity equation, followed by the balance equations for water vapor and internal energy. The last equation is the equation of state. The variables u, v, ρ, q and T are prognostic variables; p and w are diagnostic variables.

The terms $\mathcal{F}_u$, $\mathcal{F}_v$, $\mathcal{G}_q$ and $\mathcal{G}_T$ consist of the divergences of the subgridscale eddy fluxes (discussed in Sect. A.6). $\mathcal{G}_q$ includes in addition a source/sink term representing phase transitions (discussed in Sect. A.5.1) and $\mathcal{G}_T$ includes in addition a source/sink term representing the release/gain of latent heat by phase transitions and the divergence of the radiative flux (discussed in Sect. A.5.2).

These equations still contain sound waves. The hydrostatic approximation only eliminates the vertical propagation of sound waves. To filter out sound waves completely one makes the anelastic approximation. There are different ways to implement this approximation, depending on the "vertical" coordinate used. For the equations 4.16 which use z or height coordinates one usually does the following: one first introduces a dynamically irrelevant reference state $q_r(z)$, $T_r(z)$, $\rho_r(z)$ with associated hydrostatically balanced pressure $p_r(z)$ and only considers the deviation from this reference state; then one replaces the density ρ by the reference density ρ_r in all terms on the left hand side; then, most crucially, one neglects the time rate of change term in the continuity equation which then reduces to

$$w\frac{d\rho_r}{dz} + \rho_r(\nabla \cdot \mathbf{v}) = 0 \tag{4.17}$$

and becomes a diagnostic equation. This procedure eliminates sound waves completely. Only four prognostic variables, u, v, q and T, remain. The anelastic approximation is justified for motion whose phase speeds are much smaller than the sound speed, which is 330 m/s in the atmosphere and 1440 m/s in the ocean.

In the ocean, the density varies only little within the water column. One can thus replace the reference density ρ_r by a constant density ρ_0. The continuity equation then reduces to the "incompressibility" condition $\nabla \cdot \mathbf{v} = 0$. Mass conservation is replaced by volume conservation. The anelastic approximation together with this approximation is referred to as the Boussinesq approximation in oceanography. Explicitly the primitive equations in the Boussinesq approximation are given by

$$\rho_0 \left(\frac{Du}{Dt} - \frac{uv}{a}\tan\varphi - fv \right) = -\frac{1}{a\cos\varphi}\partial_\theta p' + \mathcal{F}_u$$

$$\rho_0 \left(\frac{Dv}{Dt} + \frac{u^2}{a}\tan\varphi + fu \right) = -\frac{1}{a}\partial_\varphi p' + \mathcal{F}_v$$

$$0 = -\partial_z p' - g\rho'$$

$$0 = \partial_z w + \nabla_h \cdot \mathbf{v}_h \tag{4.18}$$

$$\rho_o \frac{DS}{Dt} = \mathcal{G}_S$$

$$\rho_o c_p \frac{DT}{Dt} = \mathcal{G}_T$$

$$\rho' = \rho(S, T, p_r(z)) - \rho(S_r(z), T_r(z), p_r(z))$$

where primes denote deviations from the reference state. Note that the density ρ' is now a function of the two prognostic variables T and S and the independent variable z, through the specified reference state.

In principle, the equations 4.18 can be solved as follows: Given the values of the prognostic variables u, v, T and S at a time step one first obtains the density ρ' from the equation of state. The vertical velocity w can be calculated by integrating the incompressibility condition vertically upward using the kinematic bottom boundary condition

$$w = \boldsymbol{u}_h \cdot \nabla_h h_b$$

where h_b is the bottom elevation. The value of the vertical velocity at the surface determines the surface elevation η via

$$\frac{\partial}{\partial t} \eta = w$$

from which the the pressure at the surface can be determined via

$$p = \rho_0 g \eta$$

This value provides the necessary boundary condition to calculate the pressure p' by integrating the hydrostatic balance downwards.

The actual algorithms may differ significantly when a different "vertical" coordinate is used and when additional approximations, such as the rigid lid approximation, are applied.

4.2.4 Fundamental Cognitive Models

As we have already discussed, the understanding of the climate system has grown over the centuries. Until the appearance of powerful electronic computers in the 1950s, all progress was made through idealized cognitive models. One such model dates back to the 17th century. It is George Hadley's explanation of the trade wind system in the tropics depicted in Fig. 1.9: Air ascending in tropical convection moves poleward. Because of earth's spherical form, an air particle dislocated from a position over the equator to an off-equatorial has excess angular momentum, causing the particle to move in an eastward direction. The return flow at the surface must thus be in a southwestward direction, north of the equator, and a northwestward direction, south of the equator. This is the trade wind system.

Another important step towards comprehending climate was the understanding that the temperature at the surface of the earth is an equilibrium of incoming solar radiation and outgoing thermal radiation (see Fig. 1.11). This conceptual model is the *energy balance model*, also already discussed. When reduced to its bare essentials, it reduces to the simple model sketched in Fig. 4.9. The role of gaseous substances in modifying the radiative properties of the atmosphere, especially of carbon dioxide, had already been understood by Svante Arrhenius in the 1890s [3].

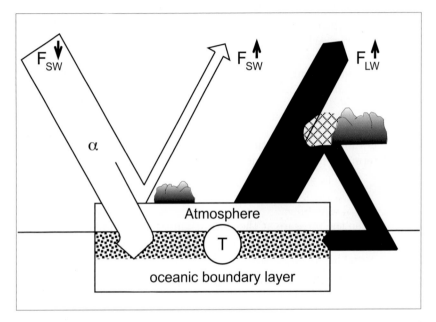

Fig. 4.9. Energy Balance Model. F_{SW} stands for short wave radiation, and F_{LW} for long wave radiation. From von Storch et al. [177]

In the early 20th century, scientists began to unravel the structure of extratropical storms. Carl-Gustaf Rossby and his colleagues discovered that these storms may be understood as unstable waves forming along the "front" separating warm and moist air masses originating from the tropics and cool and dry air masses formed in polar areas[8]. Other important conceptual models were developed for the characteristics of the Hadley Cell [60], the general circulation of the atmosphere [102], the oceanic wind-driven gyre circulations, the western intensification of these gyres [119], [160], the conveyor belt, and the oceanic mixed layer.

---

[8] The expression "front" was coined by the Norwegian pioneers in synoptic meteorology, Vilhelm Bjerknes and coworkers, under the impression of the First World War (cf. [41]).

4.2.5 Natural and Anthropogenic Climate Variability

Climate changes all the time (Fig. 4.10). This fact has not really been appreciated until recently. Climate was considered unchanging, in the sense that one cool season would eventually be balanced by a warm season, in a few years time. Even after the detection of massive prehistoric glaciation events, meteorologists still maintained the view that climate does not change within different historical time periods. This view became opposed by geographers and hydrologists who claimed that anthropogenic and other systematic climate changes are always ongoing. One widespread concern was for example that the earth would dry out, caused or at least accelerated by deforestation. For an historical account of these developments see [93] or [155].

Nowadays, climate research acknowledges the presence of two competing processes in the observational record and in any reasonable scenario of expected future climate evolution. These two processes are *anthropogenic climate change* and *natural climate variability*. They have similar signatures, namely low-frequency climatic modifications, and are therefore sometimes confused.

"Climate Variability" arises from "natural" mechanisms unrelated to human actions. One distinguishes between external and internal natural variability.

Natural external variability is caused by external forcings such as volcanism, variations in the energy output of the sun or the celestial Milankovitch cycles. "Internal variability", on the other hand, is the low-frequency variability that arises from dynamical processes within the climate system.

There are basically two internal processes, which pump energy into low-frequency variability: non-linear interactions (e.g., [69]) and the accumulation of short time-scale "weather noise" [58] (see review by Frankignoul [39]), [29]. These mechanisms make climate a particularly complex system, with properties significantly different from most classical systems.

Figure 4.10 displays two time series, describing climate variations during historical times, derived from historical accounts. Both diagrams present a mix of external and internal natural variability. One diagram shows the temperature and wetness of winter half years 1496–1995 in Switzerland [133], the other diagram the severity of ice conditions in 1500–1995 in the western Baltic sea [89].

The Swiss winters have been classified as either warm or cold, wet, dry or normal. In the diagram, the number of winter months having been classified as cold, cold and wet, or cold and dry are shown as negative values, and the number of winter months classified as warm, warm and wet, or warm and dry as positive values. The time series is characterized by variability on all time scales. The decade-to-decade variations are, according to present knowledge, mostly due to internal climate dynamics, but the outstanding cold two decades at the end of the 17th century, named *Late Maunder Minimum* [104], [194], are likely related to an external event, namely a combination of reduced solar output and the extensive presence of volcanic aerosol in the strato- and

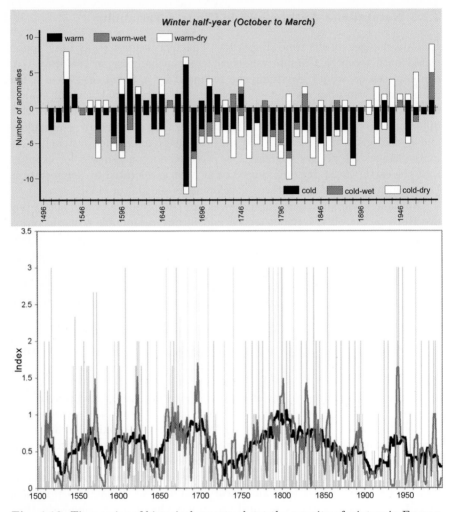

Fig. 4.10. Time series of historical reports about the severity of winters in Europe. *Top:* Number of winter months in northern Switzerland per decade reported as cold or warm (and being at the same time wet or dry) [133]. *Bottom:* Severity of ice conditions during the winters 1500–1995 in the Western Baltic sea. *Vertical lines* give annual values; the curves represent smoothed values. From Koslowski and Glaser [89]

troposphere [196] (see also Sect. 6.1.2). Also, the extended cooling from the late 15th century to the middle of the 19th century may be related to factors like the sun. However, the entire low-frequency variability cannot be ascribed to external causes alone; instead a significant proportion is related to internal variability. Thus, one cannot expect identical climatic states for the same external forcings. Generally, the shorter the time scale, the less important the external influence and the more important the internal dynamics.

A similar conclusion about the coldness of winters in central Europe is suggested by the second time series in Fig. 4.10. It is based on data that are totally independent from those used in the analysis of the Swiss winters. Again, strong variations (of sea ice cover) from year to year are present, but after smoothing the curve, clusters of colder and warmer winters emerge. Again, the Late Maunder Minimum episode stands out, as well as the generally cool time until the middle of the 19th century.

The term "Climate Change" is reserved to denote the formation of persistent climatic anomalies that are due to human activities. Examples of such activities (see, for example, Cotton's and Pielke's monograph, [19]) are urbanization (i.e., the modification of surface fluxes in the boundary layer through buildings and streets), desertification and deforestation, and anthropogenic emissions of soot (Gulf war in 1991; [5]), aerosols and greenhouse gases. The emission of CO_2, CFCs and methane, in particular, have been the subject of widespread concern because of their implications for Global Warming.

Though the potential of humankind interfering with climate has had and still has profound social and psychological implications, from a dynamical point of view it is just another forcing of the climate system.

5

Modeling in Applied Environmental Sciences – Forecasting, Analysis and Scenarios

In this and the following chapter, we consider applications of quasi-realistic environmental models. Chapter 5 deals with applied research, and Chap. 6 with fundamental research. Of course the border line between applied and fundamental research is far from rigid. For us, applied environmental research is driven by the quest for knowledge about the present and future state of a system. Thus forecasting and scenario building are in the realm of applied research. Fundamental environmental research, on the other hand, is driven by the quest for understanding a system. How does a system function? What are its significant constituents? Thus derivation and confirmation of hypotheses are typical applications in fundamental research. Obviously not all scientific applications of quasi-realistic models may be consistently classified into these two categories (see Fig. 1.13).

When computer models are used in applied environmental sciences, the main three purposes are (i) the determination of the state of the environmental system ("analysis"), (ii) the forecast of its state in the (near) future, and (iii) the description of the consequences of anthropogenic interferences. Among the secondary applications are the calibration of models by fitting them to observed data, the design of observational networks, and the determination of model uncertainties. Model uncertainties are, for instance, relevant for discriminating between natural and anthropogenic climate change and for assessing the significance of extreme values, say of wave climates. We discuss these applications in this chapter, beginning with forecasting (Sect. 5.1), data analysis (Sect. 5.2) and scenarios (Sect. 5.3), and then address secondary applications in Sect. 5.4.

5.1 Operational Forecasts

5.1.1 Forecast Versus Prediction

We do not use the two terms *forecast* and *prediction* interchangeably. Instead, forecasts refer specifically to efforts to estimate unconditionally the state of the system in some future. The term prediction is used in a broader sense; it covers forecasts of the future, but it also covers efforts to specify the outcome of an experiment, in "what-if"-analyses. Also forecasts conditional upon certain external factors, like elevated atmospheric levels of greenhouse gases, are in that sense "predictions".

In general terms, forecasts are estimates of future states and their evolution. Examples are the maximum temperature tomorrow in Copenhagen, the timing of the next high tide at Honolulu, the seasonal mean temperature of the contiguous USA during the next winter, or the timing and abundance of next year's spring algae bloom in the North Sea. These types of forecasts have economic and behavioral utility. It does not matter why the forecast is correct, but only that it produces reliable statements useful for the planning of the economic, public or personal life. To be useful, forecasts must be specific in terms of time, location and variable. Traditional long-term weather forecasts as, e.g., published in the "Old Farmers Almanac" in the contemporary USA or as asserted by the once famous Rudolf Falb in 19th century Austria [14], are often in vague terms: "There will be thunderstorms in late August". Such statements are useless simply because they are trivial or entirely random in their success [100].

Forecasts may be "point" forecasts, e.g., the next high tide will be 30 cm above the mean high tide, or "probabilistic", e.g., the next high tide will be consistent with a certain probability distribution like a uniform distribution between 25 and 30 cm or a Gaussian distribution with a mean of 30 cm and a standard deviation of 10 cm. Sometimes, categorical forecasts are also phrased in probabilistic terms as in "the probability for next summer's temperatures to be above normal is 80%" [182][1].

Some *predictions* are actually conditional forecasts as in "if the concentration of greenhouse gases in the atmosphere is increased by about 1% each year, then the global mean temperature will rise by about 3°C in the next hundred years". Such predictions are often named "scenarios" as they provide plausible evolutions based on some assumptions which may, or may not, prove to be valid. We come back to scenarios in Sect. 5.3.

Predictions are also a tool of fundamental research. An example is "whenever the stratification becomes unstable, a vertical exchange of mass takes place". This prediction can be used to test the fluid dynamical theory of convection. The classical approach for assessing a scientific theory is to contest it by checking its ability to make correct predictions. If all predictions derived from a hypothesis are found to be correct, or at least non-conflicting with observational evidence, then the hypothesis is accepted for the time being and considered a validated theory, maybe even a law [171]. We deal with this aspect in detail in Chap. 6.

---

[1] This phrase is to be interpreted that in 80% of the cases when conditions like those at the time of the forecast prevail, temperatures will be above normal.

In Sects. 5.1.2–4 we present examples of operational forecasts, with cases related to tide and storm surge, weather, and ENSO forecasts. In Sect. 5.1.5 we address the problem of how to determine the quality or skill of forecast schemes. Post-processing forecasts are discussed in Sect. 5.1.6.

5.1.2 Tide and Storm Surge Forecasts

Forecasts of tides were originally prepared by harmonic analysis of a tidal record and extrapolating the sinusoidal partial tides into the future. At the beginning of the last century, this was done with sophisticated analog machines like the one shown in Fig. 1.3. Nowadays, this method is still in use but complemented by computer models that integrate Laplace tidal equations forward in time, with or without the assimilation of tidal observations. Because of the deterministic character, such forecasts can be made for long lead times. Annual forecasts are provided in the form of annual tide tables (Gezeitentafeln). The Bundesamt für Seeschiffahrt und Hydrographie in Hamburg provides upon request "tidal predictions for all places where the data required for computation are available, and for any period of time".

For short lead times, information like

Amrum Wittdün

tide	high	low	high	low
day	26.12.2000	26.12.2000	27.12.2000	27.12.2000
time	13:54	20:25	02:06	08:46
deviation from mean	$-0,20\,\mathrm{m}$	$-0,20\,\mathrm{m}$	$\pm0,00\,\mathrm{m}$	$-0,10\,\mathrm{m}$

mean high tide: 6.21 m, mean low tide: 3.54 m

is offered regularly on the Internet for various locations.

An *operational forecast system* for various environmental variables describing the state of the North Sea and the Baltic Sea is routinely used by the Bundesamt für Seeschiffahrt und Hydrographie in Germany [28]. The system is composed of a number of modules (Fig. 5.1) that deal with currents, water levels, water temperatures, salinities, ice coverage and the drift and dispersion of substances. The system uses as input tidal predictions, meteorological and ocean wave forecasts provided by the German Weather Service, and produces forecasts 48 hours in advance. The forecasts are used by the storm surge warning service. The system also prepares drift forecasts for oil, chemicals, and floating objects, forecasts of current and water levels as a service to shipping and emergency agencies, and water quality analyses.

As an example, Fig. 5.2 shows the forecast of salinity in the North Sea and Baltic Sea. An overview of the state of the art is given by Flather [37].

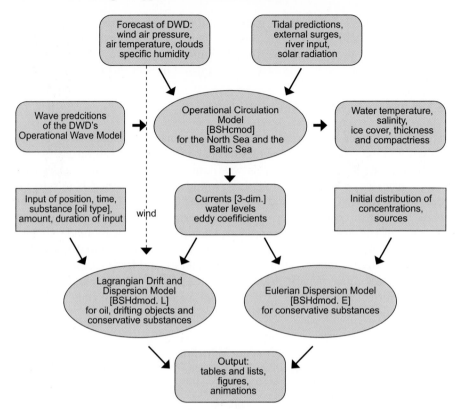

Fig. 5.1. Bundesamt für Seeschiffahrt und Hydrographie-forecast model system. It combines the processing of data and forecasts (of weather etc.) with dynamical models of the North Sea and Baltic Sea. Courtesy of Bundesamt für Seeschiffahrt und Hydrographie

5.1.3 Weather Forecasts

Numerical weather predictions (NWP) are routinely prepared by national weather services throughout the world. The ingredients are a dynamical model of the troposphere (named NWP model), a few specific parameters (like seasonal insolation), an initial distribution of the relevant meteorological variables (like temperature, wind velocities, humidity) and boundary conditions (like sea surface temperature or vegetation). Most models are global, but regional models are also in use.

Serious attempts to predict future weather by applying dynamical equations had already been made at the beginning of the 20th century. The most famous attempt, and failure, was made by Lewis Fry Richardson[2] [140]. Richardson used a simplified version of Bjerknes's primitive equations. He prepared

---

[2] He used an inadequate numerical scheme, see Appendix B.2.2.

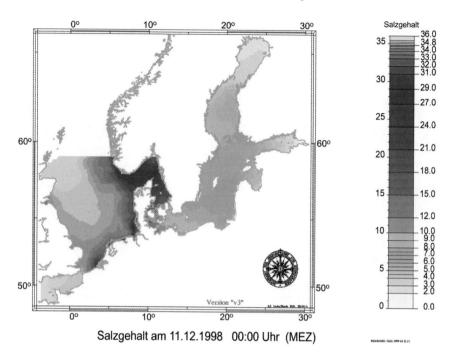

Fig. 5.2. Salinity distribution in the North Sea and Baltic Sea as predicted by the Bundesamt für Seeschiffahrt und Hydrographie forecast system, for December 11, 1998. Courtesy of Bundesamt für Seeschiffahrt und Hydrographie

within six weeks one eight-hour forecast. Even before Richardson, the Austrian Felix M. Exner [35] reported a series of 4-hourly and 12-hourly forecasts, which he prepared for the contiguous United States (see [114]):

> Exner assumes that the atmospheric flow is geostrophically balanced and that the thermal forcing is constant in time. He uses observed temperature values to deduce a mean zonal wind. He then derives a prediction equation representing advection of the pressure pattern with constant westerly speed, modified by the effects of diabatic heating.

In Fig. 5.3 one of Exner's 4-hour forecasts of the surface pressure changes is reprinted together with the corresponding observed changes, demonstrating some remarkable non-trivial skill of the forecast.

NWP with computer codes based on dynamical equations began in the late 1940s with experiments at the Institute for Advanced Studies in Princeton, when John von Neumann and Jule Charney began implementing the forward integration of dynamical equations on a digital computer [17].

This method was first implemented for operational forecasts in the 1950s at the International Meteorological Institute in Stockholm, under the leadership

Fig. 1.

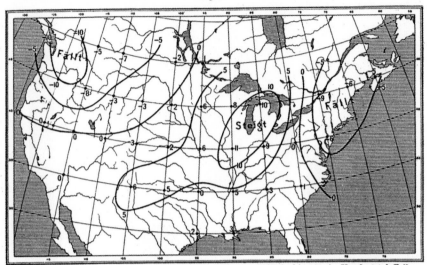

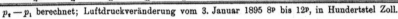$p_2 - p_1$ berechnet; Luftdruckveränderung vom 3. Januar 1895 8ᴾ bis 12ᴾ, in Hundertstel Zoll.

Fig. 2.

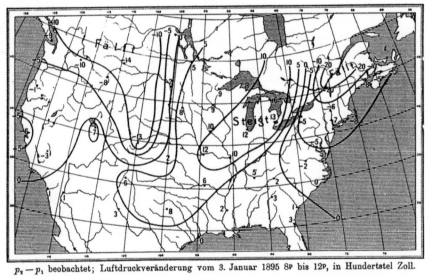

$p_2 - p_1$ beobachtet; Luftdruckveränderung vom 3. Januar 1895 8ᴾ bis 12ᴾ, in Hundertstel Zoll.

Fig. 5.3. Exner's weather prediction: 4-hour pressure changes as predicted (*top*) and observed (*bottom*), in units of 0.01 inch [35]

500 hPa   1 October 1954   0300Z

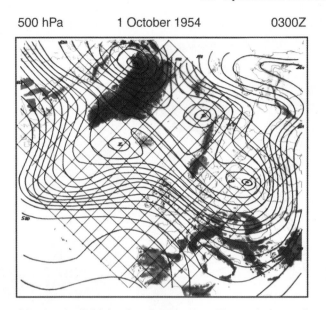

Fig. 5.4. 500 hPa height field for October 1, 1954. The grid shows the rectangular integration area with a grid distance of 300 km. The field is used as initial state for the forecast shown in Fig. 5.5. From Bengtsson [7]

of Carl-Gustaf Rossby. The original model run in 1954 made use of only one equation, the barotropic vorticity equation [154]. A rectangular area covering Northern Europe (see map in Fig. 5.4) was overlaid by a grid with a mesh size of 300 km. The time step was 1 hour. The computer was a "BESK". It had a memory of 512 words of 40 bits each. The preparation of a 1-day forecast needed about 30 minutes and 3×10^7 operations. The forecasts were extended over 3 days. The day 1 forecast was considered useful. Figure 5.4 shows the initial conditions on October 1, 1954 and Fig. 5.5 the three day forecast for October 4, 1954, together with the analysis for the same day. The forecast is not terribly good, but also not entirely useless. More on the early history of numerical weather prediction can be found in [123] and [116].

These initial numerical models evolved into operational weather forecasting as we know it today. Progress along the way was due to various factors [7]. One important factor was the increase in computer power. In 1998, the Fujitsu VPP 700 at the European Center for Medium Range Forecasts integrated 200-variables at 31 vertical levels with a horizontal resolution of about 50 km (T213). One time step was 20 min. It executed 5×10^{12} operations per forecast day in about 6 minutes. It computed complete 3-dimensional distributions for each time step and each variable. The forecasts were skillful for 6–9 days in advance. Other important factors were better observations, improved dynamical equations, better numerical methods, increased resolution,

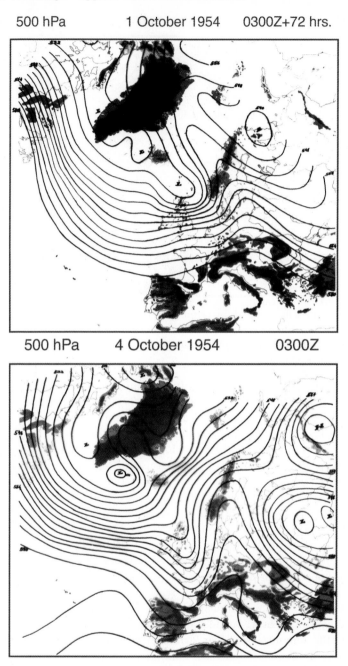

Fig. 5.5. 500 hPa heights for October 4, 1954. *Top:* barotropic forecast, initiated with the field shown in Fig. 5.4. *Bottom:* analysis. From Bengtsson [7]

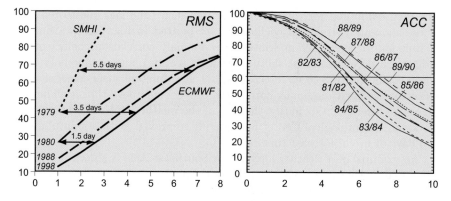

Fig. 5.6. Development of the skill of weather forecast systems for the Northern Hemisphere during winter. *Left:* Root mean square error (RMS) of weather forecasts prepared by SMHI and ECMWF since 1979 as a function of forecast day. After Bengtsson [7]. *Right:* Anomaly correlation coefficients (ACC) of observations and forecasts prepared by the US National Meteorological Center in 1981/82 to 1989/90 as a function of time lag in days. Operational weather forecasters usually consider 60% as a threshold for useful forecasts. From Kalnay et al. [75]

advanced physical parameterizations and a better incorporation of observational evidence through data assimilation[3].

The gain in forecast skill is displayed in the left part of Fig. 5.6. It shows the root mean square error of the 1000 hPa height field forecast[4], averaged over the Northern Hemisphere and many cases[5] as a function of days forecasted. The curve "1979" shows the skill of the model of the Swedish weather service SMHI in 1979. It reaches an error of over 40 m within one day and shows strong error growth thereafter. The other curves show the performance of the ECMWF model in 1980, 1988 and 1998. The typical error of the SMHI forecast of about 40 m for a 1-day forecast was reached by ECMWF with its 1980 model after about 2 days, with its 1988 model after about 3.5 days and with its 1998 model after 4 days. Obviously, great progress has been made, but this progress seems to level off.

A similar success has been achieved by the National Meteorological Center of the US Weather Service, as demonstrated in the right part of Fig. 5.6. 10-day forecasts of the Northern Hemisphere 500 hPa height field in the winters 1981/82 to 1989/90 were analyzed. The anomaly correlation coefficient[6]

---

[3] For a discussion of data assimilation methods, refer to Sect. 3.2 and Appendix D.

[4] The spatial distribution of the height at which the atmospheric pressure is 1000 hPa.

[5] For a more detailed discussion of skill scores, refer to Sect. 5.1.5 and Appendix C.2.4.

[6] The anomaly correlation coefficient is the area averaged correlation between the anomalies of the predicted and observed field.

was then computed for each winter and for each time lag, one to ten days. These anomaly correlation coefficients are plotted in Fig. 5.6 as a function of time lag for the different winters. An anomaly correlation coefficient of 1 indicates a perfect forecast skill (apart from scaling), and an anomaly correlation coefficient of zero no forecast skill at all. Meteorologists consider 0.6 as a critical level, with anomaly correlation coefficients above that level indicating useful forecasts. Obviously, the skill of the weather forecast scheme of the National Meteorological Center has steadily improved, with a gain of almost two days of useful forecast time.

An example of a weather forecast for Southeast Asia is shown in Fig. 5.7. The target is shown as "ANA", which is the analyzed state on September 12, 2003 (a randomly chosen date). Also shown are the forecasts for that target date, issued 24 hours (1 day), 48 hours (2 days), 72 hours (3 days), 120 hours (5 days) and 192 hours (8 days) earlier. These forecast were initialized with analyses like the one shown as "ANA".

The forecast issued 1 day earlier is very good; for instance, it predicts the presence of a small cut-off low south of the main North Pacific low pressure system. This small scale phenomenon is not foreseen 3 and more days earlier. A closer scrutiny of the maps reveals that the skill in forecasting details deteriorates when the forecast lead times becomes longer. After 8 days the similarity between forecast and observed state stems from the common late summer climatology, such as the tendency of the pressure to be low over the North Pacific and high over China. A randomly chosen map from the same season would have a comparable similarity with the observed state. This is a general result of such NWP models. They do a credible forecast for lead times of a few days, but lose their skill after 8 or so days.

A significant practical problem is the preparation of the initial condition for NWP models. It cannot be observed in its entirety. Only isolated point measurements of most variables are available, plus satellite images for a few spatially distributed variables. In a first step, a weather analysis scheme (see Sect. 5.2) combines this information with, or "assimilates" it into, the most recent forecast. This step results in a consistent, complete analysis of the 3-dimensional weather state. In a second step, this state is "initialized", i.e., modes of motion not represented in the NWP model are eliminated from the initial state[7].

Predicting weather and weather statistics for many months in advance has been a dream of meteorologists from the beginning. The seminal findings of Lorenz [102] put a damper on this dream. Chaotic systems can only be

---

[7] At any given time, the atmosphere is made up of a mix of various modes of motion. Some of these modes are relevant for weather forecasting, like Rossby waves, while others are irrelevant, like sound waves and meteorological tides. These irrelevant modes are not described by weather forecast models. They would require very short time steps. The dynamical equations are therefore filtered to suppress these modes of motion (see Sect. A.9.1). These modes must also be eliminated from the initial state for dynamical consistency. We come back to this issue in Sect. 5.2.

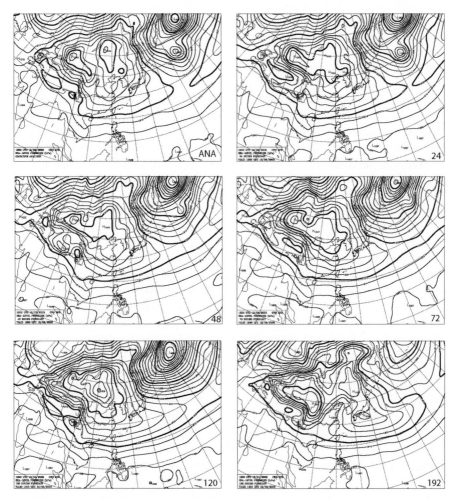

Fig. 5.7. Spatial distribution of air pressure at the surface as analyzed on December 9, 2003 ("ANA") and as predicted for that day 24, 48, 72, 120 and 192 hours in advance. Courtesy Central Weather Bureau, Taiwan

predicted for a limited time, and mid-latitude weather is the most prominent example of such a chaotic system. Efforts are underway for the preparation of seasonal forecasts for mid-latitude areas by establishing ensembles of extended forecasts. In the tropics seasonal forecasts are feasible utilizing the prolonged persistence of weather anomalies in the tropics. This case is discussed in the next section.

5.1.4 El Niño Southern Oscillation Forecasts

Weather anomalies in the tropics persist for much longer times than in mid-latitudes and become part of climate. To understand their dynamics, and thus their predictability, the international research program TOGA 1985–95 (Tropical Ocean–Global Atmosphere) was initiated. Its main finding was that tropical climate variations, and their impact on extratropical areas, must be understood as a result of interactions of the ocean and the atmosphere. El Niño/Southern Oscillation (ENSO) was identified as the most important mode of tropical variation. It is the major mode of natural climate variability on the interannual time scale. Its formation is due to complex atmosphere–ocean interactions in the Tropical Pacific. Its memory extends over many months and originates from the oceanic inertia.

More specifically, ENSO describes the following mode of variation. It had been discovered at the end of the last century [62], [185] that sea-level pressure (SLP) in the Indonesian region is negatively correlated with that in the southeast Tropical Pacific. A positive SLP anomaly (i.e., a deviation from the long-term mean) over, say, Darwin (Northern Australia) tends to be associated with a negative SLP anomaly over Papeete (Tahiti). This seesaw pattern is called the "Southern Oscillation" (SO). The SO is associated with large-scale and persistent anomalies of sea surface temperature in the central and eastern Tropical Pacific called "El Niño" and "La Niña". The combined phenomenon is referred to as "El Niño/Southern Oscillation" (ENSO). The ENSO phenomenon is also associated with large zonal displacements of the centers of precipitation. These displacements reflect anomalies in the location and intensity of the meridionally (i.e., north–south) oriented Hadley Cell and of the zonally (east–west) oriented Walker Cell.

The state of the Southern Oscillation is conventionally monitored by the monthly SLP difference between observations taken at surface stations in Darwin, Australia and Papeete, Tahiti. This difference is called the Southern Oscillation Index (SOI). Figure 5.8 shows a time series of the SOI since the 1930s. There are also many other ways to define equivalent SO indices [189]. Areal averages of sea surface temperature (SST) are very common. An example of such a SST based index is also shown in Fig. 5.8. It is highly correlated with the pressure based SOI. The most frequently SST-index is the so-called *Niño 3 index*, which is the sea surface temperature averaged over the area $150°–90°W \times 5°S–5°N$ in the central Tropical Pacific. It will be used below.

The observed SO indices are rather persistent, though the persistence has a marked annual cycle, as displayed in Fig. 5.9. During January and February (*vertical axis*), the lag correlations decay quickly with lag (*horizontal axis*), while in March and April moderate correlations extend for almost a year. Later in the year, for instance in September, correlations are high for several months, but then drop quickly for longer lead times.

The ENSO phenomenon was found to be predictable first with empirical methods and then later with dynamical models, for a limited time. As an

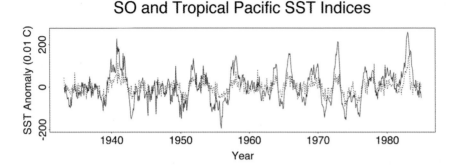

SO and Tropical Pacific SST Indices

Fig. 5.8. The conventional Southern Oscillation Index (SOI = pressure difference between Darwin and Tahiti; *dashed curve*) and a sea-surface temperature (SST) index of the Southern Oscillation (*solid curve*) plotted as a function of time. From von Storch and Zwiers [182]

example, we consider the dynamical coupled atmosphere–ocean model used by the Center for Ocean–Land–Atmosphere (COLA) [83]. The model consists of an atmospheric and an oceanic component. The atmospheric component is a global spectral atmospheric general circulation (AGCM) model triangularly truncated at total wave number 30 with 18 unevenly spaced levels in the vertical. The ocean model is a Pacific basin version of the Geophysical Fluid Dynamics Laboratory (GFDL) ocean model with 1.5° longitude by 0.5° latitude resolution in the tropics and 20 levels in the vertical. The oceanic and atmospheric components are *anomaly coupled*. Only the predicted anomalies of wind stress and SST are exchanged interactively at the air–sea interface.

An important task is the initialization of the model, i.e., the preparation of the initial state of the coupled model. Since the atmosphere is largely in equilibrium with the ocean it is the ocean that needs to be initialized. The COLA model uses an iterative procedure that is designed to reduce simulated sea surface temperature anomaly (SSTA) errors in the eastern Pacific [83]. Neither data from the TOGA TAO array of buoys in the Tropical Pacific nor from the TOPEX/Poseidon satellite are assimilated. In fact, the inclusion of such in-situ data for initializing the forecast may bear significant potential for improving the overall forecast system[8].

Figures 5.10–15 demonstrate the skill of the COLA model to forecast both horizontal distributions as well as the characteristic Niño-3 index. Figures 5.10 and 5.11 display the predicted and observed winter (DJF) sea surface temperature anomalies for the El Niño event 1997/98 and the La Niña event 1998/99. The prediction lead time is 6 to 9 months in advance, i.e., the hindcasts[9] were

---

[8] Such systems are prepared at NCEP and ECMWF.

[9] A retrospective forecast, or hindcast, is a forecast prepared for a certain time in the past, using only data before that time. The advantage of this procedure is

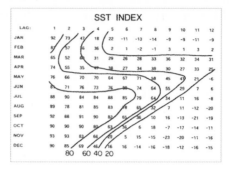

Fig. 5.9. Seasonal dependence of the lag correlations of a SST index of the Southern Oscillation. The correlations are given in units of 0.01 so that isolines represent lag correlations of 0.8, 0.6, 0.4, and 0.2. The row labelled 'Jan' lists the correlations between the January values of the index and the values of the index observed 'n lag' months later [190]

prepared with data available in May 1997 and May 1998, respectively. In both cases, the hindcast is successful in the Tropical Pacific – which is the main objective of the system in the first place – whereas outside that region the similarity between hindcasted and observed SSTA is low. The broad warming along the equator in 1997/98 as well the cooling in the winter 1998/99 are well reproduced by the dynamical model, even though details, as for instance the conditions off the Peruvian coast, are less well captured. The intensity of the warming and cooling of the central Tropical Pacific is also well reproduced.

The COLA model has also been used to hindcast sea surface temperature anomalies in the Tropical Pacific since the mid 1960s. Figure 5.12 shows Niño-3 indices, as observed and as hindcasted (+) with a lead time of 9 months. Obviously, the model is quite successful, even though it tends to underestimate large values of the indices, for instance in the 91–93 El Niño event. The correlation between the hindcast, with different lead times, and the observed Niño-3 index is displayed in Fig. 5.13, in the same format as Fig. 5.9, but with the base months on the vertical axis inverted. Similarly to the observed seasonal characteristics of the persistence of the Southern Oscillation Index, the model exhibits maximum forecast skill in March, with remarkable correlations for lags up to 12 months, while forecasts issued in October show high correlations only for lead times of up to 5 or 6 months in advance[10].

---

that forecasts can be done for many cases and statistics of the forecast skill be obtained. Researchers usually evaluate different forecast schemes by hindcasting. If the number of cases is not very large – as in the case of ENSO – the outcome of this comparison may depend on particularities of the considered cases.

[10] Figure 5.13 is based on fewer numbers than Fig. 5.9. Thus, Fig. 5.13 is subject to irregular sample variations. This may explain the large differences between neighboring months, like in November or January. It appears plausible that the curves should be smooth, as in Fig. 5.9.

Observed SSTA DJF97−98

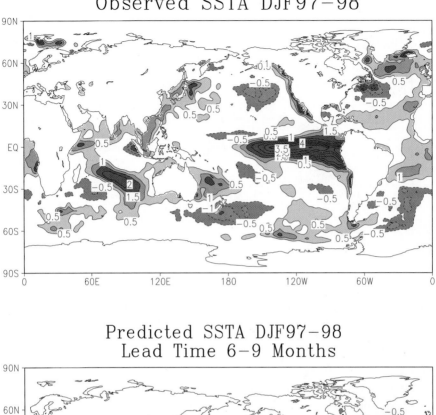

Predicted SSTA DJF97−98
Lead Time 6−9 Months

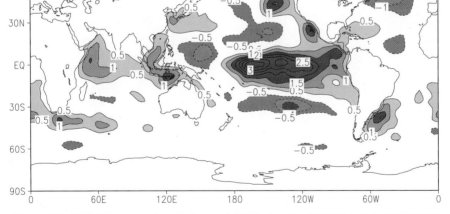

Fig. 5.10. COLA-forecasted and observed SSTA for the "El Niño winter" 1997/98. The forecast lead time is 6 to 9 months, i.e., the forecast was prepared with data available in May 1997. Courtesy: Ben Kirtman

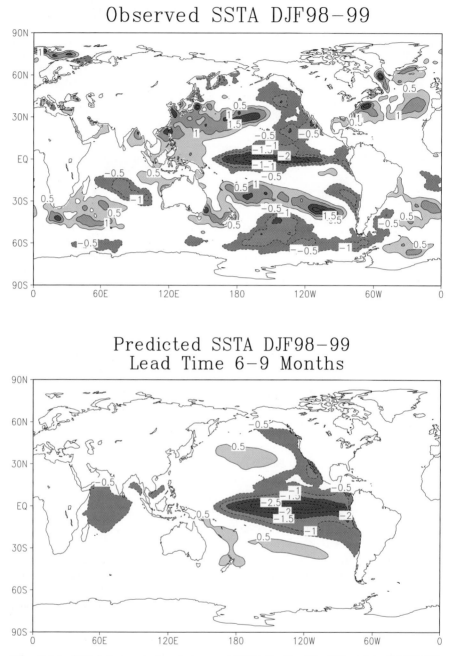

Fig. 5.11. COLA-forecasted and observed SSTA for the "La Niña winter" 1998/99. The forecast lead time is 6 to 9 months, i.e., the forecast was prepared with data available in May 1998. Courtesy: Ben Kirtman

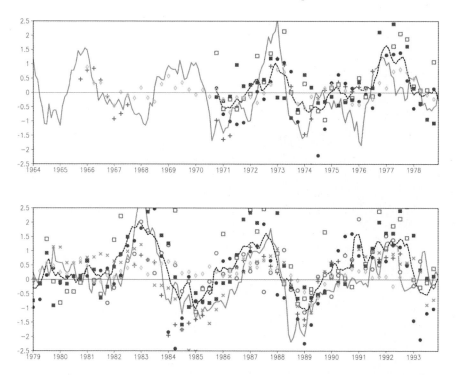

Fig. 5.12. 9-months hindcasts of Niño-3 index with a total of 6 different models, the average of the six hindcasts ("consensus" model; *dashed*) and the observations to be replicated (*solid*). The COLA model hindcasts are given by crosses +. From Kirtman et al. [82]

Kirtman et al. [82] assessed the state-of-the-art in predicting Niño-3 anomalies with several dynamical atmosphere–ocean models. Hindcasts made by various research groups with five different dynamical models and one empirical scheme were compared. One of the models is the above mentioned COLA model. A seventh model, named *consensus model*, is the average of the forecasts of all 6 models.

The performance of the hindcasts is compared in terms of several *skill scores*. One such skill score is the correlation between hindcasted and observed Niño-3. These correlations are given in Table 5.1 for the COLA model, the NCEP model (which assimilates TOGA–TAO buoy data) and the consensus model, and for different lead times.

From long series of hindcasts, confidence limits for the correlations can be determined. When serial correlations are disregarded, then a separation of 0.05 is sufficient to conclude, with a reasonable risk of 5%, that the correlations are different. One would thus conclude from Table 5.1 that the consensus model is best for 6 and 12 months lead times, and that COLA fares better than NCEP

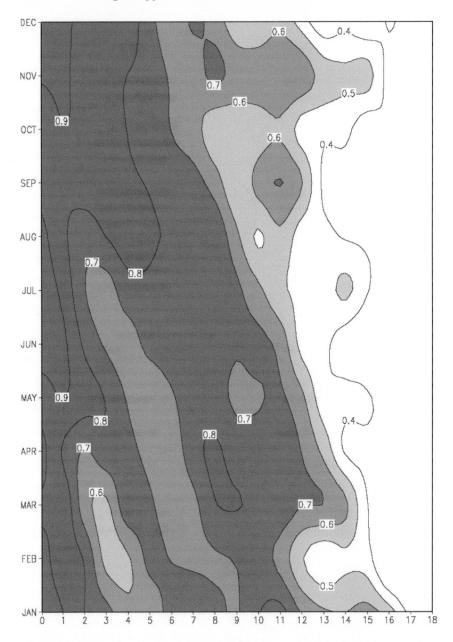

Fig. 5.13. Correlation between the observed and the COLA-hindcasted Niño-3 index. The correlation is computed for different base months (*vertical axis*) and for different monthly lead times n (*horizontal axis*). For instance, the value for January and $n = 3$ represents the correlation between Niño-3 indices in April as observed and as forecasted 3 months earlier in the previous January. From Kirtman et al. [82]

Table 5.1. Correlations between hindcasted and observed Niño-3 indices for the COLA, NCEP and consensus models

	6 months	9 months	12 months
COLA	0.70	0.67	0.56
NCEP	0.75	0.66	0.50
consensus	0.77	0.67	0.63

for the two longer lead times, while NCEP is more successful for the shorter lead time of 6 months.

Comparisons have also been made by Landsea and Knaff [94] to determine how well the models forecast the onset and the decay of ENSO. They studied in detail how well various empirical and dynamical models[11] forecast the very strong warm event in 1997/98. They concluded that none of the considered models adequately captured the detailed life cycle of this very strong ENSO event[12].

As an example, Fig. 5.14 shows the 6-months lead time forecasts of two relatively successful models, the COLA and NCEP models, together with the observed indices[13]. The COLA model fails to anticipate the emergence of the event. Only in northern spring (FMA), when the event was already beginning to develop, did this model predict a (much too small) warming for late summer (ASO). The model's forecast of the event ending in early summer 1998 (MJJ) turned out, however, to be correct. On the other hand, NCEP was successful in expecting already in fall (SON) of 1996 an onset in MAM 1997, but incorrectly envisaged in the winter (DJF) 1998 that the event would extend well into 1998. Other models failed to predict the event altogether: onset, peak and decay.

---

[11] Landsea and Knaff examined all forecasts routinely published in the *Experimental Long-Lead Forecast Bulletin*, http://www.iges.org/ellfb, with lead times of at least 2 seasons in advance (6–8 months). One of the models was a "consensus" model, i.e., a weighted average of several other models, but it did not fare very well, contrary to the finding of Kirtman et al. [82]. Also the advanced model of the ECMWF was not included in the comparative analysis, because forecasts with sufficiently long lead times were not made publicly available (Landsea, pers. comm.).

[12] Of course, the analysis of only one, albeit very strong, event limits the conclusions that can be drawn from this study. However, this event was for most models the only event for which data did definitely not enter the design of the models. Also, one could argue that a scheme that fails for this very strong event will likely not be very successful for more common (i.e., less intense) El Niño-events.

[13] Note that the two models predicted different indices. COLA predicted the Niño-3 index whereas NCEP predicted the so-called Niño-3.4 index, which is rather similar but not equal to Niño-3. Also, the models used different 3-months intervals for seasons. COLA used, for instance, FMA = February, March and April, whereas NCEP used MAM = March, April and May.

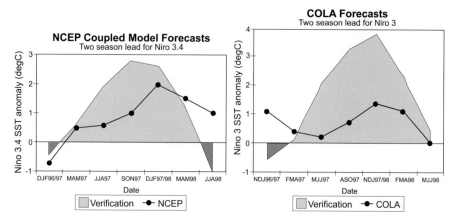

Fig. 5.14. The strong El Niño-event in 1997/98. The observed Niño indices are given in red. The *black dots* are 6 months lead time forecasts of the Niño indices prepared by the NCEP and COLA model. These two models were found among the most successful in predicting the life cycle of the 1997/87 El Niño-event [94]

We come back to these forecasts, when we deal with the problem of measuring the skill of forecasts in the next section.

As in case of weather forecasting, the value of an ENSO-forecast will greatly increase if its skill can be estimated. One approach for quantifying the skill is to try to relate the consistency of forecasts initialized one month apart to the error of the forecasts. Unfortunately, this approach has failed so far, mainly because the persistence of the phenomenon makes consecutive forecasts relatively consistent, even if they are grossly wrong. Better ensemble techniques need to be developed which describe the sensitivity of the forecast to small errors [82].

In conclusion, both statistical and dynamical ENSO models produce useful tropical SSTA forecasts for the peak phase of ENSO up to two seasons in advance [82], while early indicators for the onset and decay of ENSO events are not very reliable [94]. It may very well be that forecasting ENSO may hit insurmountable barriers, as is the case for weather forecasting, because the ENSO dynamics is inherently chaotic. It may also very well be that not all available information has been utilized yet. In particular, there is some optimism that the implementation of efficient methods to assimilate local observations, from the TOGA–TAO buoy array and from satellites recording sea level and sea surface temperature, will advance the skill of the forecast significantly.

5.1.5 The Skill of Forecasts

Here we elaborate on the particular concept of skill that we have already used in our discussion of ENSO and of weather forecasting. The task of defining fair measures of the skill of a forecast system is far from trivial. In the following we present the general concepts and give examples. Some technical details are given in Appendix C.2.4. A detailed discussion can be found in [100] and [182].

The most straightforward manner to assess the skill of a forecast is to make many forecasts and to count the number of successes. This approach requires each time a decision whether a forecast is correct or not. This is no problem for *categorical* forecasts. Categorial forecasts predict a finite number of discrete events. Examples are "above normal" or "below or equal to normal". *Point forecasts*, on the other hand, predict a specific number out of a continuum, say a 20 cm water level, but the probability that the water level eventually observed is *exactly* 20 cm, and not 20,003 cm, is nil.

As discussed in Sect. 3.1.2, quantitative measures of skill regard the forecast **F** and the corresponding observations, the predictand **P**, as the two components of a bivariate random variable. For categorical forecasts one can explicitly examine the joint probability distribution of the forecast **F** and the predictand **P** in a contingency table. For continuous forecasts the examination of the joint probability distribution becomes very elaborate and one extracts indices, called skill scores, that characterize the joint probability distribution and the performance of the forecast. The most common skill scores are the correlation skill score, the mean square error, and the Brier skill score.

Example 5.1. As an illustration of a categorical forecast we consider the 24-hour forecast of the temperature in Minneapolis (Minnesota) during winter. The bivariate random variable (forecast temperature, observed temperature) are binned into boxes of $1\,F \times 1\,F$ in order to transform the continuous forecast into a categorical forecast. The *top panel* of Fig. 5.15 shows the relative number of entries for these boxes. "Correct" or "near-correct" forecasts are indicated by open circles for better identification. This figure represents the joint probability (relative frequency) distribution of the bivariate random variable (**F**,**P**). The highest relative frequencies correspond mostly to correct forecasts, but for forecasts $\mathbf{F} \leq 28°F$ the predictands tend to be systematically lower than the forecast by a few degrees.

The conditional distribution of the predictand given the forecast is shown in the bottom panel of Fig. 5.15 where the 10%, 25%, 50%, 75%, and 90% quantiles of the observations are plotted against the forecast temperature. The forecast is *conditionally unbiased* if the solid curve, representing the conditional 50% quantile, lies on the diagonal. This is not the case. In particular, the mean observed temperature is about 3°F lower when temperatures below 20°F are forecast. When temperatures below 12°F are forecast, about 75% of the observations are actually lower than the forecast.

The conditional standard deviations of the forecast errors are of the order of 5°F, and forecast errors larger than 20°F never occur. Very little can be learned about the skill of forecasts below 8°F and above 48°F because of poor sampling.

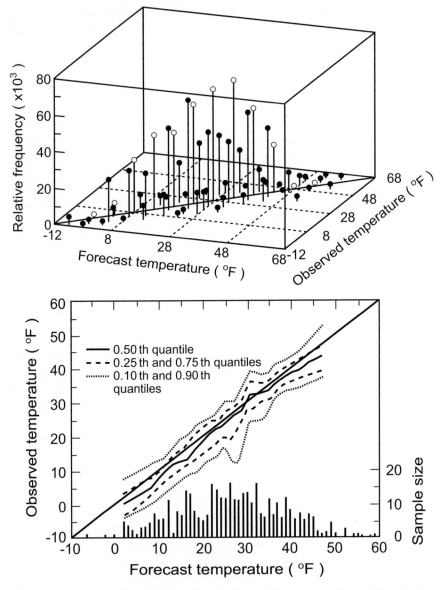

Fig. 5.15. *Top:* Estimated joint distribution of forecasts and predictands for a 24-hour temperature forecast in Minneapolis during winter. Temperature is given in *Fahrenheit.* All data are collected into small bins. Cases with correct forecasts are indicated by open circles to facilitate identification. *Bottom:* Quantiles of the distribution of the predictand conditional on the forecast. The number of the forecasts is also shown so that the credibility of the conditional quantiles can be assessed. From Murphy et al. [121]

In the above example we have considered a skill of a forecast scheme without reference to other schemes. Thus we learn something about the scheme, but we do not learn if it is worth the effort, or if a similar forecast skill could be obtained by a simpler scheme. Such a *reference scheme* is often called a *strawman scheme*. The next example demonstrates this concept.

Example 5.2. As an example of a continuous forecast we consider the Brier skill score (C.35) for a number of ENSO forecasts. The Brier skill score compares the performance of the forecast **F** relative to a reference forecast **R**. A Brier skill score larger than zero indicates that the forecast is better than the reference. A Brier skill score smaller than zero indicates the opposite. In many cases, the reference forecast **R** is simply persistence. The forecast scheme is required to "beat" persistence, in order to qualify as a useful scheme. However, in some cases more complex reference forecasts are in use. Landsea and Knaff [94] compared a series of dynamical ENSO-forecast models against their own much simpler empirical scheme ENSO-CLIPPER. The performance of their own model during the strong 1997/98 El Niño-event is displayed in the inset of Fig. 5.16. The Brier skill scores of various other dynamical ENSO forecast models are shown in the main figure. Their own "strawman" model performs rather well. It is therefore not surprising to see that the Brier skill scores of the other models are mostly below zero, indicating that they do not do better than the reference model. The COLA model (*filled circles*) reaches a slightly positive skill for a lead time of three seasons. The NCEP model (*stars*) remains just below zero within the considered lead time of two seasons. In spite of public announcements "Models win big in forecasting El Niño" in *Science* 1998, dynamical models still have to make headway to outcompete economically less demanding empirical models [94]. Progress should be expected from improved data assimilation schemes, which will help the dynamical models to better capture the initial evolution of ENSO.

Many applications do not give single forecast numbers, such as the temperature in Minneapolis on a certain day or the Niño-3 index for a certain month, but spatial distributions, such as weather maps or distributions of sea surface temperature anomalies in the Tropical Pacific. For assessing the quality of these forecasts, one does not need many realizations or samples but can instead use area averages. Examples are the root mean square error between the forecasted and observed fields or the area averaged correlation between the two fields. Usually, anomalies are considered and the correlation becomes the *anomaly correlation coefficient*. When forecasts are prepared for extended times, then time series of the root mean square error or the anomaly correlation coefficient may be derived, and episodes with better and worse performance of the forecast scheme may be identified. Such root mean square errors and anomaly correlation coefficients were given in Fig. 5.6 which demonstrated the improvement of the skill of weather forecasting.

Even if dynamical forecast models have skill, they will in most cases suffer from some systematic errors. The forecasts may be biased so that the

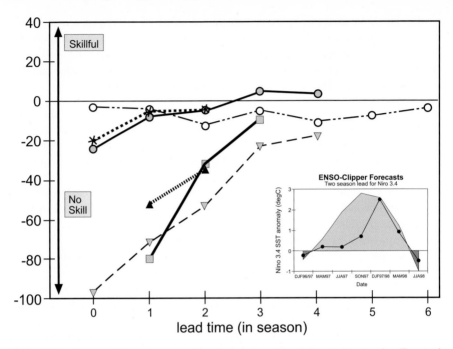

Fig. 5.16. Brier skill score of various forecasts of an Niño-index in the Tropical Pacific, using the CLIPPER forecast shown in the inset as a strawman. Forecasts with scores above zero are considered as more skillful than the strawman. From Landsea and Knaff [94]

predictand is systematically under- or overpredicted. Also the variance of the forecast may be systematically larger or smaller than the observation. In these cases it is common practice to apply a simple correction, namely to subtract the bias and to rescale the forecast.

5.1.6 Post-processing Forecasts

Forecast models often have difficulties in accurately computing local values, especially if these values are affected by details of the local physiography (coast lines, valleys, forests). Also, users often ask for variables, such as sunshine hours, that are not computed by NWP models. In these cases, a method called "Model Output Statistics" (MOS) is applied. MOS is so widely used that it is referenced in the Encyclopædia Brittanica:

> Another technique for objective short-range forecasting is called MOS (for Model Output Statistics). Conceived by Harry R. Glahn and D.A. Lowry of the U.S. National Weather Service, this method involves the use of data relating to past weather phenomena and developments to extrapolate the values of certain weather elements, usual forecasts for

a specific location and time period. It overcomes the weaknesses of numerical models by developing statistical relations between model forecasts and observed weather. These relations are then used to translate the model forecasts directly to specific weather forecasts. For example, a numerical model might not predict the occurrence of surface winds at all, and whatever winds it did predict might always be too strong. MOS relations can automatically correct for errors in wind speed and produce quite accurate forecasts of wind occurrence at a specific point, such as Heathrow Airport near London. As long as numerical weather prediction models are imperfect, there may be many uses for the MOS technique.

In most cases, the MOS model is formulated as a multiple regression problem [48], [85]:

$$\omega = \sum_k a_k \psi_k + \delta \tag{5.1}$$

with a predictand ω, predictors ψ_k and an unexplained part δ. The predictors ψ_k in (5.1) are variables forecasted by the numerical model, such as geopotential height and humidity. The predictand ω is the desired meteorological variable, like sunshine hours recorded at a specific location. Thus, (5.1) combines simulated ψ_k values with observed ω values. It is a special case of an observation equation as introduced in Sect. 3.2. The regression coefficients a_k are fitted to a large ensemble of simultaneous ω and ψ_k values. Then, given a NWP forecast with ψ_k, the best forecast for the local value ω is

$$\hat{\omega} = \sum_k a_k \psi_k \tag{5.2}$$

A distinct disadvantage of the MOS regression model (5.1) is its dependency on the specific model version. When the model is changed, and this happens regularly in the routine of weather services, then the regression coefficients a_k need to be fitted anew.

An example of the MOS methodology is the prediction of the minimum and maximum temperatures and the snow amounts at the Alta Guard Station in Utah, an observing post operated by the US Forest Service and the Utah Department of Transportation at an altitude of 8730 ft (Gibson, pers. comm.). These data are supposed to help avalanche forecasting. Here we will only consider the minimum and maximum temperature.

Data recorded regularly in the winter time from January 1992 until December 1996 were used to fit multiple regression equations of the type (5.1) to predictor variables forecasted routinely by the "Nested Grid Model" (NGM) of the National Center for Environmental Prediction (NCEP). The NGM predicts geopotential height, potential temperature, zonal and meridional wind components, relative humidity at all levels and grid points of the NGM, and in addition mean sea level pressure, the 1000–850 hPa thickness and the 850–500 hPa temperature difference. These variables are forecasted by NGM for

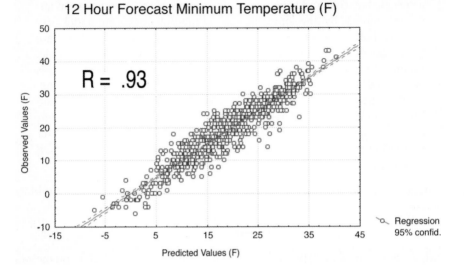

Fig. 5.17. Scatter plot of the 12 hour observed and MOS predicted minimum temperature at Alta Guard Station, Utah in *Fahrenheit*. The predictions represent 86% of the variance of the local observations. Chris Gibson, National Weather Service

6, 12, 18 etc. hours in advance. When the predictand ω is the "minimum temperature 12 hours in advance", then the variables "6 hour forecast of 1000–500 hPa thickness", "6 hour forecast of relative humidity at 700 hPa", "12 hour forecast of 750 hPa temperature" and "6 hour forecast of zonal wind at 500 hPa" are found to be the most powerful set of predictors ψ_k. These predictors describe 86% of the variance of the minimum temperature at Alta Guard Station. The performance of this fit is displayed in Fig. 5.17 as a scatter plot.

Using independent recorded data from the winter 1998, the skill of the MOS forecast was compared with the simple persistence forecast. The skill is measured by the mean absolute error $S = \frac{1}{n}\sum_j |f_j - \hat{f}_j|$, where f is the local observations and $\hat{f}$ the MOS forecast at the n winter days indexed by j. The result is summarized in Fig. 5.18. It shows the impressive improvement of MOS over persistence. For the 12 hour minimum temperature $S = 3.02\,\text{F}$ for MOS, but $S = 5.55\,\text{F}$ for persistence.

Other empirical a posteriori processing schemes are in use as well. The MOS equation (5.1) may be replaced by a nonlinear scheme, as for instance a neural net formulation, as long as enough data are available to train the net. Another frequent approach is "downscaling" which purports to derive dynamically consistent local details from simulation runs with climate models. An *empirical* model is constructed that relates observed large-scale fields to the observed local variables of interest. The empirical model is then applied to the output of a climate model. Downscaling presumes that the regional

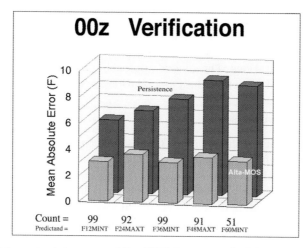

Fig. 5.18. Mean absolute errors of the MOS forecast and of the persistence forecast for minimum (MINT) and maximum (MAXT) temperature 12 hours (F12), 24 hours (F24) etc. in advance. Units are *Fahrenheit*. Chris Gibson, National Weather Service

climate results from the interaction of large-scale climate features (such as the intensity and location of the jet-stream) with the regional physiographic details (such as the topography and land–sea distribution). Details can be found in [174]. Downscaling is widely used, in particular, when regional and local details of climate change scenarios are sought.

5.2 Data Analysis

The seemingly simple task of describing the present, and possibly recent past, state of the environment is far from trivial. There is a series of problems. One challenge is to make reliable, reproducible measurements, for instance of the concentration of a substance, the abundance of a species, the transport of sediment or the precipitation over the ocean. This problem is not considered here. A second problem is that the environment is neither uniform nor homogeneous but subject to variations on all time and space scales. The relevant scales of atmospheric, oceanic and coastal modes of motion were sketched in Fig. 4.7 and Fig. 4.8. Thus, any point observation is only representative for a short time interval and a small volume. To obtain complete fields these point observations have to be interpolated. Dynamical models can increase the information content of such interpolated fields by making them dynamically consistent. The method of combining (limited) observations and (dynamical) models, for whatever purpose, is called *data assimilation*. The data are *assimilated into the models*. When the purpose is a dynamically consistent field estimate the method is called *data analysis* by meteorologists. Data analysis is also applied to satellite data, which cover the ocean surface, to infer information about the state of the ocean below the surface. In the following, we present some examples of such model-supported "analyses" and discuss their potentials and limitations. The general approach has already been sketched in Sect. 3.2. Technical aspects of data assimilation are described in Appendix D.

5.2.1 Global Reanalyses

The state of the atmosphere cannot be observed in its entirety. The various land-, ship-, aircraft- and satellite-based observing systems provide point observations and vertical profiles irregularly spaced in the atmosphere, together with distributions on the land and ocean surfaces. These data are used by operational weather centers to construct, or "analyze", continuous distributions of atmospheric variables. Such "analyses" are our best guess of the atmospheric state and deviate from the true, unknown state to some extent. The large scales are better described, simply because they are better sampled. Details on scales of a few tens of kilometers and less are insufficiently sampled and subject to significant uncertainty.

In the beginning of modern weather forecasting, the analyses were prepared by hand. A major first insight of meteorology was the finding by the Bergen school at about 1920 that the appearance of the sky contains information about the state of the atmosphere that should be incorporated into the analysis and weather forecasting process [41]. The advent of satellites in the 1970s, with their quasi-complete mapping of horizontal distributions, caused another major improvement. The final breakthrough was the systematic interpretation of observational data aided by quasi-realistic dynamical models.

In the 1990s, the National Center for Environmental Prediction (NCEP) in cooperation with the National Center for Atmospheric Research (NCAR) and the European Center for Medium Range Forecasts (ECMWF) made significant separate efforts to analyze meteorological data of the past with modern modeling and data assimilation methods. The results were the so-called *re-analyses*. These re-analyses provide complete 4-dimensional data sets. These data sets are not completely *homogenous*. They are free of changes in the weather statistics due to changing analysis tools but not to changing observational practices and changing density of observations (such as the introduction of satellites in the 1970s; see Fig. 5.19). These data sets have become widely used both by the atmospheric science community as well as in climate monitoring studies.

In the following the NCEP/NCAR project will be discussed, following the presentations by Kalnay et al. [75] and Kistler et al. [84]. This project resulted in a 51-year (1948–1998) record of global analyses of atmospheric fields. It first collected and quality controlled data from weather stations, ships, rawinsondes, pibals[14], aircrafts, satellites and other observational platforms since 1948. Specifically it used:

---

[14] Pibal = pilot balloons = balloons tracked visually or with radar.

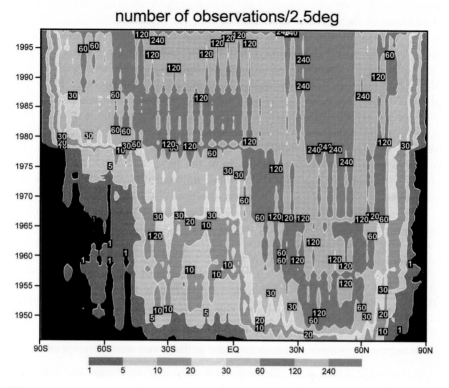

Fig. 5.19. Number of all observations per 2.5° latitude and per months from 1946 until 1998. 12-months running mean [84]

- upper air rawinsonde observations of temperature, horizontal wind and specific humidity,
- operational vertical temperature soundings over the ocean from polar orbiting satellites,
- temperature soundings over land above 100 hPa,
- cloud tracked winds from geostationary satellites,
- aircraft observations of wind and temperature,
- land surface reports of surface pressure, and
- oceanic reports of surface pressure, temperature, horizontal wind and specific humidity.

The number of these data has considerably increased since the late 1940s, as is displayed in Fig. 5.19. A significant increase took place during the International Geophysical Year in 1957, when the modern rawinsonde network was established. Since 1979, satellite data were routinely included. Clearly, the quality of the re-analyses in the early years cannot be as good as in the later years. Especially the re-analyses of the earliest decade (1948–1957), when observations were few and primarily in the Northern Hemisphere, are less reliable than those of the last four decades.

As the second step all these data were assimilated into a state-of-the-art NCEP global spectral weather prediction model. The forecasting model is kept unchanged ("frozen"). The system continues to be used with current data, so that re-analyses are available from 1948 to the present. The result of the NCEP project is a relatively homogenous complete history of global weather maps. The maps have a horizontal gridding of about 210 km (T62 spectral truncation) and 19 pressure-levels in the vertical. The assimilation method used the forecast model as a weak constraint[15] (see Appendix D). The resulting maps are thus not a solution of the model. Especially, they do not necessarily conserve quantities like mass, momentum and energy. This has, however, turned out not to be a severe problem.

The re-analyses generate many variables. Some of these depend strongly on the observational data, while others depend more strongly on the forecast model and its parameters. To discriminate among these different cases, four categories ("flags") "A" to "D" were introduced:

A: Analysis is based strongly on observed data.
B: Analysis is partially based on observed data but significantly influenced by the model characteristics.
C: Analysis is based on model alone.
D: Analysis represents climatologies and is fixed and model-independent.

At 17 pressure levels, (1000, 925, 850, 700, 600, 500, 400, 300, 250, 200, 150, 100, 70, 50, 30, 20, 10 hPa) the following variables are given on a 2.5° latitude–longitude grid (144 × 73 grid points) every 6 hours:

A: Geopotential height (gpm), u-wind (m/s), v-wind (m/s), temperature (K), absolute vorticity (1/s)
B: Pressure vertical velocity (Pa/s), relative humidity (%)

As two-dimensional grids, the following long list of variables have been stored every 6 hours:

A: Temperature at the tropopause (K), pressure at the tropopause (Pa), u-wind at the tropopause (m/s), v-wind at the tropopause (m/s), vertical speed shear at the tropopause (1/s), temperature at the maximum wind level (K), pressure at the maximum wind level (Pa), u-wind at the maximum wind level (m/s), v-wind at the maximum wind level (m/s), pressure reduced to MSL (Pa), pressure at the surface (Pa).

B: Precipitable water (kg/m²), relative humidity of the total atmospheric column (%), relative humidity in 3 sigma layers: 0.44–0.72, 0.72–0.94, 0.44–1.0 (%), potential temperature at the lowest sigma level (K), temperature at the lowest sigma level (K), pressure vertical velocity at the lowest sigma

---

[15] The model is considered imperfect, and its predictions are deemed uncertain.

level (Pa/s), relative humidity at the lowest sigma level (%), u-wind at the lowest sigma level (m/s), v-wind at the lowest sigma level (m/s), maximum temperature at 2 m (K), minimum temperature at 2 m (K), specific humidity at 2 m (kg/kg), temperature at 2 m (K), zonal component of momentum flux (N/m^2), u-wind at 10 m (m/s), v-wind at 10 m (m/s).

C: Cloud forcing net longwave flux at the top of atmosphere (W/m^2), cloud forcing net longwave flux at the surface (W/m^2), cloud forcing net longwave flux for total atmospheric column (W/m^2), cloud forcing net solar flux at the top of the atmosphere (W/m^2), cloud forcing net solar flux at the surface (W/m^2), cloud forcing net solar flux for total atmospheric column (W/m^2), convective precipitation rate (kg/m^2/s), clear sky downward longwave flux at the surface (W/m^2), clear sky downward solar flux at the surface (W/m^2), clear sky upward longwave flux at the top of the atmosphere (W/m^2), clear sky upward solar flux at the top of atmosphere (W/m^2), clear sky upward solar flux at the surface (W/m^2), cloud work function (J/Kg), downward longwave radiation flux at the surface (W/m^2), downward solar radiation flux at the top of the atmosphere (W/m^2), downward solar radiation flux at the surface (W/m^2), ground heat flux (W/m^2), latent heat flux (W/m^2), near IR beam downward solar flux at the surface (W/m^2), near IR diffuse downward solar flux at the surface (W/m^2), potential evaporation rate (W/m^2), precipitation rate (kg/m^2/s), pressure at high cloud top (Pa), pressure at high cloud base (Pa), pressure at middle cloud top (Pa), pressure at middle cloud base (Pa), pressure at low cloud top (Pa), pressure at low cloud base (Pa), pressure at the surface (Pa), run off (kg/m^2 per 6 hour interval), nearby model level of high cloud top (integer), nearby model level of high cloud base (integer), nearby model level of middle cloud top (integer), nearby model level of middle cloud base (integer), nearby model level of low cloud top (integer), nearby model level of low cloud base (integer), sensible heat flux (W/m^2), volumetric soil moisture content (fraction) (2 layers), total cloud cover of high cloud layer (%), total cloud cover of middle cloud layer (%), total cloud cover of low cloud layer (%), temperature of the soil layer (3 layers) (K), temperature of high cloud top (K), temperature of low cloud top (K), temperature of middle cloud top (K), zonal gravity wave stress (N/m^2), upward longwave radiation flux at the top of the atmosphere (W/m^2), upward longwave radiation flux at the surface (W/m^2), upward solar radiation flux at the top of the atmosphere (W/m^2), upward solar radiation flux at the surface (W/m^2), meridional gravity wave stress (N/m^2), visible beam downward solar flux at the surface (W/m^2), visible diffuse downward solar flux at the surface (W/m^2), meridional component of momentum flux (N/m^2), water equivalent of accumulated snow depth (kg/m^2).

D: Ice concentration (ice = 1; no ice = 0), land–sea mask (1 = land; 0 = sea), surface roughness (m).

This complete data set can potentially be used for all kinds of studies, such as the determination of 50-year trends, of the skill of a forecast scheme and of operationally "unobservable" quantities. Major questions are, of course, how much information has been gained by the re-analyses and how suitable the re-analyses are for such studies. These issues are addressed next:

Trends

Meaningful trends cannot be determined from the re-analysis data sets, because the density of the observations is inhomogeneous in time (cf. Fig. 5.19). For the same reason, low-frequency variations cannot be determined.

Forecast performance

The information content of the re-analysis can be assessed by the *observational increments*. These are the differences between a forecast initialized with the re-analysis and the re-analysis at the time of the forecast. To facilitate such an assessment the NCEP project has also produced 5-day "re-forecasts" throughout the re-analysis period. In the following three examples, the "observed" state is not really observed but a re-analysis itself.

Figure 5.20 shows, as a skill score, the anomaly correlation coefficient between the 5-day NCEP re-forecasts and the re-analyses at the time of the forecast, separately for the Northern and Southern Hemisphere. The first thing to notice is that the forecast skill is not constant, but increases steadily. This indicates that the quality, or the degree of realism, of the re-analysis has steadily improved due to better and more observations. Not surprisingly, the re-analysis is much better on the Northern Hemisphere, where many more observations are available, than on the much less sampled Southern Hemisphere. Note that the high skill scores at the beginning of the analysis period, prior to the International Geophysical Year 1957, are artificial. They are due to the fact that hardly any data were available on the Southern Hemisphere at that time with the result that the analyses, to which the forecasts were compared, were essentially the forecasts themselves. Another interesting detail is the fact that prior to about 1990 the forecasts initialized with the re-analyses performed better than the forecasts initialized with then state-of-the-art analysis systems.

A second example is the forecast of the catastrophic storm in January/February 1953 which caused great damage in the Netherlands. Figure 5.21 shows a four-day forecast of the pressure distribution on 1 February 1953, initialized with the re-analysis of January 28. The forecast compares well with both the re-analysis of that day and the analysis prepared manually by contemporary Dutch meteorologists. At the time, the event could not have been predicted with a reasonable advance warning time, since neither realistic dynamical forecast models nor data assimilation methods were available; the re-forecast prepared by NCEP demonstrates though that the *data* available in 1953 were sufficient for a good forecast. Obviously, atmospheric sciences have made significant progress since the 1950s, not only in terms of understanding but also in terms of practical applications, such as forecasting and warning.

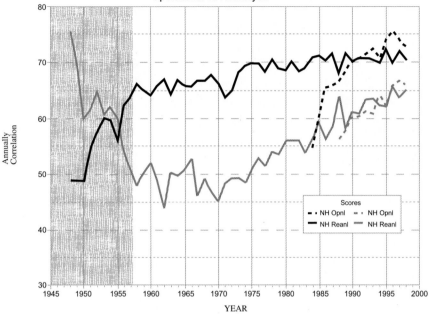

Fig. 5.20. Annually averaged anomaly correlation coefficients between the 5-day NCEP re-forecasts and the re-analysis at the time of the forecast, for the Northern (*black*) and Southern Hemispheres (*grey*). The *dashed curves* are the anomaly correlation coefficients of the operational forecasts since the 1980s. From Kistler et al. [84]

Our third example uses the re-forecasts to assess the relative role of observational systems. Figure 5.22 displays the time mean anomaly correlation coefficients for the year 1979, separately in the Northern and Southern Hemisphere. The forecasts were initialized once with re-analysis states that included satellite data, and once with re-analysis states that excluded satellite data. "Reality" for both forecasts was the re-analysis with satellite data. On the Northern Hemisphere, the inclusion of satellite data had little effect. On the Southern Hemisphere, on the other hand, the incorporation of satellite information into the initial fields greatly improved the performance of the forecast. Obviously, Southern Hemisphere analyses from the pre-satellite period should be considered with care.

"Unobservable" quantities

The "C" variables of the NCEP re-analysis are completely determined by the model. They contain many near-surface variables such as precipitation

96 hour SLP forecast for Feb 01 03Z 1953

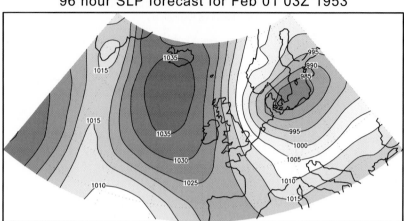

SLP digitized handanalysis for Feb 01 03Z 1953

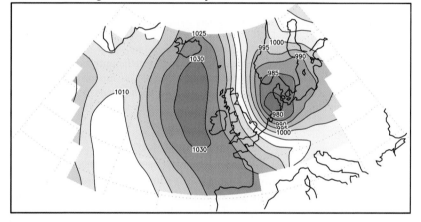

Fig. 5.21. Forecast and analysis of the storm from January 31 to February 2, 1953, which caused catastrophic flooding in the Netherlands. The *top* shows a 4-day forecast for 03Z, February 1, 1953, initialized with the re-analysis four days earlier. The *bottom* shows a digitized version of a contemporary manually prepared analysis for 00Z, February 1, 1953 (van den Dool, pers. comm.). The contour interval is 5 hPa

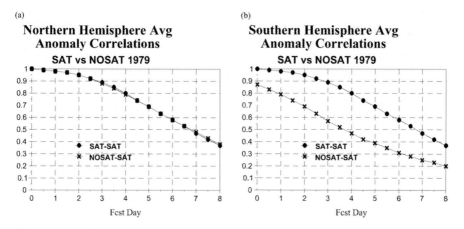

Fig. 5.22. Mean anomaly correlation coefficients between re-forecasts and re-analyses in 1979 for the Northern and Southern Hemisphere. The re-forecasts were initialized either with analyses that included satellite data (*dots*) or with analyses that excluded satellite data (*crosses*). The re-analyses that these forecasts are compared with include satellite observations [84]

and surface fluxes[16]. Comparisons of such "C" variables with independent near-surface in-situ data will thus provide powerful tests of the realism of the re-analyses. One such independent data set is the *Comprehensive Ocean Atmosphere Data Set* (COADS) of in-situ ship observations.

Figure 5.23 displays the correlation of evaporation time series from COADS with those of the re-analyses. Note that the figure fails to indicate that COADS does not provide a reasonable data coverage south of about 30°S, in the central Indian Ocean and in the central and eastern Southern Hemisphere subtropical Pacific. In the remaining areas COADS variations of evaporation are recovered to a large extent by the operational NCEP/NCAR re-analyses. Low correlations might, of course, also be due to the limitations of the COADS data set. It is reassuring that the largest correlations are found in areas with good shipping coverage in the past 50 years, like the Northern North Atlantic.

Table 5.2 also supports the notion that the "C" variables of the re-analyses contain useful information about the real world. It lists time mean averages of the heat flux over the ocean surface from the NCEP/NCAR re-analyses, the ECMWF re-analyses, COADS and satellites (from [84]).

On the other hand, one must not forget that the numbers in the re-analyses are educated guesses. They will in many cases deviate systematically from real conditions. As an example consider Fig. 5.24. It shows a time series of daily

---

[16] The re-analyses are mostly determined by upper air observations and little by surface observations. This is partly due to the fact that near-surface physiographic details (such as the topography) are insufficiently represented and partly due to the fact that the observed surface variables have little dynamical impact on the upper-air state.

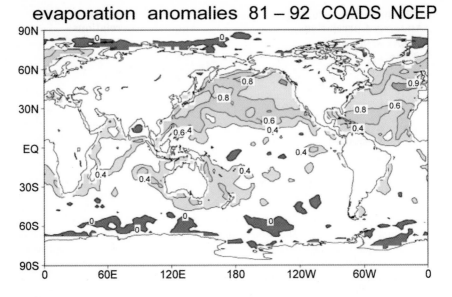

Fig. 5.23. Correlation of evaporation time series over the ocean, derived from the COADS in-situ ship records, with estimates from the NCEP/NCAR re-analyses. Note (i) that the COADS data did not enter the re-analysis process, and (ii) that NCEP considers evaporation a C-variable, i.e., a variable that is not directly affected by the assimilation process, but mostly determined by the forecast model [84]

Table 5.2. Time mean averages of the heat flux over the ocean surface from different re-analyses and observations

heat flux (W/m^2)	COADS	ECMWF	NCEP	satellite
sensible	−10	−9.8	−10.9	−
latent	−88	−103	−93	−
net shortwave	170	160	166	173
net longwave	−49	−50.6	−56.4	−41.9

precipitation in East Asia derived from the re-analysis and from an average across many rain gauges in that region during the summer monsoon 1991. The re-analyses exhibit too much rainfall over the whole monsoon season and too few days without rainfall.

Overall, a comparison of NCEP "C" variables with independent observations implies that they generally contain considerable useful information, especially for seasonal and interannual variability [84].

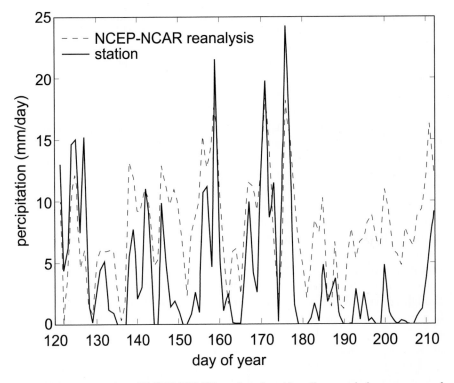

Fig. 5.24. Time series of NCEP/NCAR analyzed and locally recorded area-averaged precipitation in an area of East Asia (25°–30°N, 115°–120°E) from May to July 1991. Units are in mm/day. Courtesy: Yuqing Wang

5.2.2 Reconstruction of Regional Weather

The global analyses, discussed in the previous Sect. 5.2.1, can only reproduce features that are well resolved by the forecast model. The NCEP re-analyses have a horizontal grid spacing of 2.5°; the T106 ERA re-analyses, prepared by the European Center for Medium Range Forecasts, have a similar spacing. These re-analyses are found to be inadequate to represent, say, the surface winds over the Adriatic Sea. They represent the entire Adriatic Sea by only about 10 grid points and are unable to resolve the characteristic deformations of the wind flow due to the complex topography surrounding the Adriatic Sea [99]. Such lack of detail in the global re-analyses remains a major problem for regional applications. One way out is to use empirical downscaling schemes, as discussed in Sect. 5.1.6, to derive local details from global re-analyses. High-resolution regional models and regional weather observations provide another more substantive approach. They can be combined to give high-resolution regional analyses, which are useful for many applications. This program has not been carried out in a complete and systematic manner yet and there is no standardized set of regional re-analyses. The regional models require lateral boundary conditions. Here we first demonstrate that useful results are obtained when these are prescribed from large-scale models or analyses. Then we consider an effort to reconstruct the European weather at high resolution, using a regional model and NCEP re-analyses.

Various regional atmospheric models are in use to simulate regional atmospheric dynamics. An overview of these models can be found in [46]. These regional models are usually forced by time-varying fields, from a low-resolution model or from re-analyses, along lateral *sponge layers*[17]. Regional models forced in this way are in fact capable of reproducing small-scale details. This is demonstrated by the *Big Brother Experiment* (BBE) by Denis et al. [26]. They used a regional model with a grid spacing of about 50 km covering a large part of North America and integrated it over some time. This is the "Big Brother" (BB). Next, they coarsened the output of that run to a resolution of 300 km. These coarsened data were used to force the same regional model with the same grid spacing of 50 km in a smaller region embedded in the area covered by BB. This is the "Little Brother" (LB). The question is whether LB will reproduce the small scale features simulated by BB, even though LB is provided only with information about the coarsened state of BB in a narrow sponge zone of the smaller area. The answer is definitely positive, as is demonstrated in Fig. 5.25 which shows the relative humidity at 850 hPa after 4 days of integration in BB and LB [26].

The European weather was reconstructed at high resolution by the following procedure [36]. The regional climate model REMO [68] was integrated for 40 years, being forced by 6-hourly global NCEP re-analyses. REMO is a grid point model that integrates the discretized primitive equations in a terrain-following hybrid coordinate system. The finite differencing scheme is energy preserving. The prognostic variables are surface pressure, horizontal wind components, temperature, specific humidity and cloud water. A soil model is added to account for soil temperature and water content. The inte-

---

[17] The concept of sponge layers was introduced by Davies [23]. The phase speed of waves on a grid depends on the grid spacing (and on the discretization scheme; see Appendix B.2.3). Thus, the phase speed in the interior of a high-resolution model does not match its low-resolution prescription at the boundary. The boundary value problem becomes mathematically ill-posed and problems may emerge in the simulation.

Practice has, however, shown that the problem can be held at check by forcing the interior solution in a boundary layer, called the *sponge layer*, to remain close to the prescribed boundary values. The "interior" solution of the model, denoted Ψ, is restored, or "nudged", to the prescribed boundary value, denoted Ψ^*, by adding a restoring or nudging term $\gamma \cdot (\Psi^* - \Psi)$ to the equations. The "nudging coefficient" γ is largest at the lateral boundary and decreases towards the interior of the integration domain. When $\Psi > \Psi^*$, the restoring terms cause a decrease in Ψ. When $\Psi < \Psi^*$ it induces an increase of Ψ. The nudging coefficient has units of 1/time.

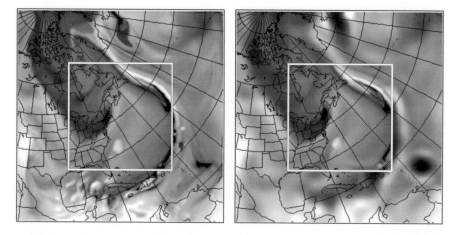

Fig. 5.25. The *Big Brother Experiment* of Denis et al. [26]. Relative humidity at 850 hPa after 4 days of integration. The full area shown is the integration domain of the "Big Brother" (BB) and the box in the interior the domain of the "Little Brother" (LB). *Left:* High-resolution representation of BB-simulation. *Right:* Outside of the box: coarsened BB-simulation; inside box: LB simulation

gration area, shown in Fig. 5.26, has a horizontal grid spacing of about 0.5° resulting in 91 × 81 grid points. A time step of 5 minutes is adopted.

In addition to the forcing by the NCEP re-analyses at the lateral boundaries and in the sponge layers, a spectral nudging technique [179] is applied to the entire model domain. It forces the model to adopt the large-scale features provided by the NCEP re-analyses, while regional features may evolve independently from this forcing. The technique adds nudging terms in the spectral domain, with maximum strengths at small wavenumbers (large wavelengths) and high vertical levels. The details of this procedure are as follows:

First the relevant REMO variables are expanded in a Fourier series

$$\Psi(\theta, \varphi, t) = \sum_{j=-J_m, k=-K_m}^{J_m, K_m} \alpha_{j,k}^m(t) e^{ij\theta/L_\theta} e^{ik\varphi/L_\varphi} \tag{5.3}$$

with zonal coordinates θ, zonal wave-numbers j, zonal extension L_θ of the model area, meridional coordinates φ, meridional wave-numbers k, and meridional extension L_φ. The number of zonal and meridional wave-numbers is J_m and K_m and the expansion coefficients are $\alpha_{j,k}^m$. The index m stands for model. A similar expansion is done for the NCEP re-analyses, which are given on a coarser grid. The number of Fourier coefficients of this expansion is $J_a < J_m$ and $K_a < K_m$ and the coefficients are $\alpha_{j,k}^a$. The index a stands for analysis. The confidence in the realism of the different scales of the re-analysis depends on the wavenumbers j and k and is denoted by $\eta_{j,k}$.

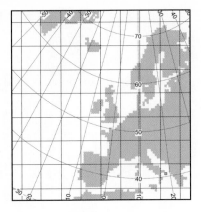

Fig. 5.26. The model area and the REMO grid. *Solid lines* demarcate 100 (10 × 10) grid boxes. Because of details of the gridding, the *straight lines* are only approximations

The REMO model is then forced to follow the state given by the re-analysis conditional upon this confidence. This is achieved by adding a "nudging term" to the equation for Ψ in the spectral domain

$$\sum_{j=-J_a,k=-K_a}^{J_a,K_a} \eta_{j,k}(\alpha_{j,k}^a(t) - \alpha_{j,k}^m(t))e^{ij\theta/L_\theta}e^{ik\varphi/L_\varphi} \qquad (5.4)$$

The higher the confidence is in the global analyses, the larger are the values of $\eta_{j,k}$, and the more efficient is the nudging term.

This nudging procedure was applied to the zonal and meridional wind components only. The nudging coefficients were set to $\eta_{j,k} = \eta^0$ for $j = 0\ldots3$ and $k = 0\ldots5$ and to $\eta_{j,k} = 0$ otherwise. Wavelengths of about 15° and larger (corresponding to 6 and more NCEP grid points) are considered to be reliably analyzed by NCEP. The nudging coefficient η^0 depends on height, so that a deviation from the NCEP re-analyses decays in about 60 days at 850 hPa, in about 1 day at 500 hPa, and in about 3 hours at the model's top level of 25 hPa. No nudging is applied below 850 hPa. This prescription allows the regional model to develop its own dynamics at the lower levels where regional geographical features are expected to be more important.

The performance of the REMO model nudged to the NCEP re-analyses in the above way is illustrated in Fig. 5.27. The *top panel* shows the modeled and observed wind speeds at 10-m height for the winter 1995 at a coastal station at the German Bight. Most characteristics of the fluctuations are well reproduced. Some of the larger peaks, for instance those of December 18 or January 28, are not reproduced by the model run. The *lower panel* of Fig. 5.27 additionally shows monthly biases and root mean square errors of the modeled wind speeds with respect to the observed wind speeds for the years 1958 to 1996. The bias shows a distinct annual cycle and a slight tendency toward

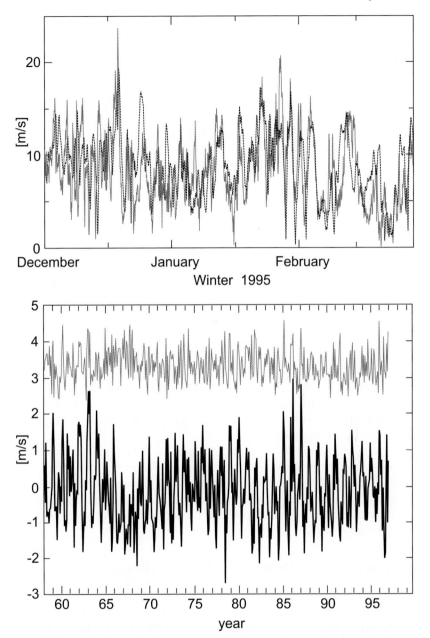

Fig. 5.27. Performance of the REMO model. *Top:* Modeled (*dashed line*) and observed (*grey line*) wind speed (m/s) at 10-m height at a coastal station at the German Bight for the winter 1995. *Bottom:* Monthly root mean square errors (*grey curve; top*) and biases (*black curve; bottom*) of the modeled wind speed at 10-m height for the period 1958–1996 [36] (© (2001) American Geophysical Union. Reproduced by permission of American Geophysical Union)

smaller values at the end of the simulation period. Overall, the true wind conditions are reproduced by the model with a typical bias ranging from -1.5 to $+1\,m/s$ and an average root-mean-square error of $3.5\,m/s$. At offshore stations, the bias and rms errors are generally much smaller than those at the coastal stations[18].

Regional models produce indeed regional details. This is confirmed by Table 5.3. It compares the variances in the zonal and meridional wind components at 850 hPa at large scales, which are well resolved by the NCEP re-analysis and thus sustained by the spectral nudging, with those at small scales, for which the regional model is supposed to add detail (from [181]).

Table 5.3. Variances of the zonal and meridional wind components at 850 hPa

	units m^2s^{-2}	NCEP analyses	REMO nudging
zonal wind			
large scale	10^{-2}	1.6	1.6
small scales	10^{-6}	3.7	8.1
meridional wind			
large scales	10^{-2}	1.4	1.5
small scales	10^{-6}	2.1	8.5

The variances at the large scales are comparable in the NCEP re-analyses and the REMO model, while the variances at the small scales, well resolved by REMO but less so by NCEP, are markedly larger in the regional model than in the global analyses.

The temporally and spatially high-resolution data obtained from this atmospheric simulation have been used as input data for a number of applied projects. One such project is the reconstruction of waves and water levels in European coastal seas[19]. Another project uses a particle transport model to examine lead transport and deposition during the last 40 years in Europe, and is discussed next.

5.2.3 Transport and Deposition of Lead in Europe

The following example treats the reconstruction of long-range transport and deposition of substances on the continental scale. It illustrates how empirical knowledge (here the daily sequence of weather events during the last 4 decades and estimates of lead emissions from, mainly, traffic) and dynamical

---

[18] As outlined in Sect. 3.1 the comparison between grid-box values and local observations is fraught with an inherent inconsistency. This inconsistency is particularly marked when the local observations are strongly influenced by local specifics as in case of a coast.

[19] HIPOCAS, http://mar.ist.utl.pt/hipocas/

knowledge (here the atmospheric transport and deposition mechanisms) may be combined to comprehensively describe environmental change and the effect of political measures. This case also serves as an example of how different data may help to validate simulations, and how simulations may be used to expand our knowledge in the case of insufficient observational data.

After decades of regulating the emission of anthropogenic substances into the environment, a retrospective analysis of the effects of such regulations is informative. It allows to determine the actual costs and benefits of the regulations. As an example we describe here the case of gasoline lead in Europe, analyzed by von Storch et al. [175]. A regional climate model, NCEP re-analyses, spatially disaggregated lead emissions from road traffic and point sources, and various local data were combined to reconstruct the airborne pathways and depositions of gasoline lead in Europe since 1958. The approach succeeded in describing the time-variable, spatially disaggregated deposition of gasoline lead.

It was discovered in the early 1920s that adding lead to fuel increases engine performance by preventing self-ignition. Lead additives, such as tetraethyl lead and tetramethyl lead, were developed. Higher-compression engines were produced and the use of leaded gasoline increased enormously all over the world. In the 1960s, increased automobile traffic resulted in air pollution problems in high-income countries. The United States reacted in 1963 to this challenge and passed the "Clean Air Act Amendment". In Europe, abatement began in the 1970s. Figure 5.28 shows gasoline sales and lead emissions from 1950 to 1992 in Germany. The gasoline sales increased substantially throughout this period. The lead emissions increased concurrently until regulations were enacted in 1972 and 1976 that drastically reduced the amount of added lead. Emissions sharply decreased. In 1985 unleaded gasoline was introduced with the effect that lead emissions dropped even further in the following years. In 1985 the EU mandated its member states to offer unleaded gas after October 1989, and recommended a maximum lead content of $0.15\,\mathrm{g\,Pb/l}$.

Spatially disaggregated estimates of lead emissions, on a 50 by 50 km^2 grid, were prepared for the years 1955, 1965, 1975, 1985, 1990 and 1995 [128] considering road traffic, metal manufacturing, stationary fuel combustion, waste disposal, cement production and other processes. The result is displayed in Fig. 5.29 for the years 1965, 1975, 1985 and 1995. The increase until 1975 due to increasing traffic and high lead concentrations in gasoline as well as the decrease afterwards, reflecting the political measures, are clearly seen in Western and Central Europe. In Eastern Europe, however, little change can be detected.

Atmospheric lead is bound to suspended matter and can be transported by wind over long distances. Thus, lead transport across Europe can be modeled with an atmospheric transport model. Lead is considered a passive atmospheric tracer, which is removed from the air by dry and wet deposition.

Lead concentrations and depositions over Europe were computed by a two-dimensional Lagrangian model. It uses 6-hourly, high-resolution weather anal-

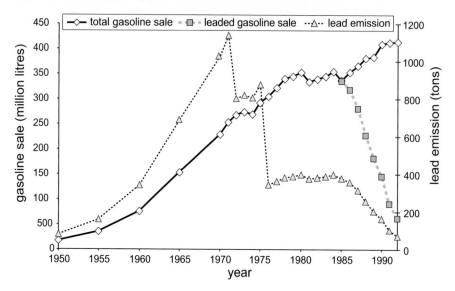

Fig. 5.28. Annual gasoline sales and lead emissions in Germany. Volume of gasoline sold (millions of liters per year; *solid*) and of leaded gas (after 1985; *grey-dashed*); amount of lead added to the gasoline (in tons; *dotted*). From Hagner [50]

yses (Sect. 5.2.2) and the lead emission estimates of Pacyna and Pacyna [128] after suitable temporal interpolation. It is assumed that lead remains within the well-mixed planetary boundary layer, where it is horizontally advected by wind and deposited to the surface through precipitation and turbulent transport. The high-resolution weather analyses were obtained by "regionalising" the NCEP re-analyses as described in the previous Sect. 5.2.2. The modeling resulted in a complete set of 6-hourly maps of lead concentrations in the planetary boundary layer covering all of Europe. Annual mean concentrations are displayed in Fig. 5.30. Not surprisingly, the annual lead concentrations in Europe increased sharply until about 1975 and diminished substantially by 1985 and 1995. Only local maxima in Southern England and the industrial areas of Belgium, Germany, Northern Italy, Ukraine and the Russian Federation remained. The simulated lead depositions (Fig. 5.31) show the same spatial and temporal patterns as the simulated lead concentrations.

To validate the model results, they were compared with local measurements of lead concentration in the air and of lead depositions. This approach is of limited value, since the observational records are fairly short and represent point observations. Nevertheless, the model output was found to be consistent with the limited empirical evidence about concentrations and depositions of lead in Europe.

Therefore the model was used for further investigations. It was assumed that the model results describe realistically the lead transports and depositions in the years 1958 to 1995. The simulated data were then analyzed

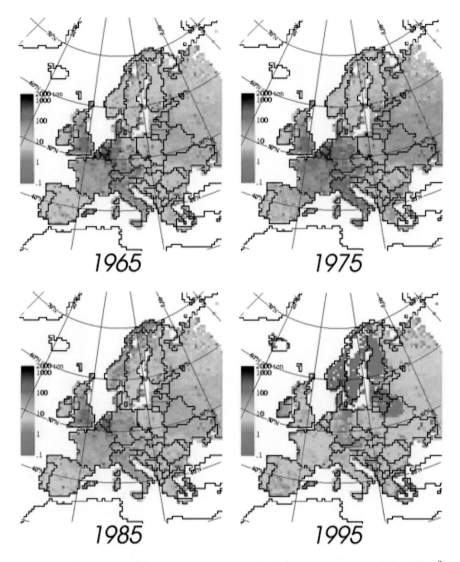

Fig. 5.29. Estimates of European emissions of lead (in tons per pixel of $50 \times 50\text{km}^2$) from road traffic and industrial processes for the years 1965, 1975, 1985 and 1995. From Pacyna and Pacyna [128]

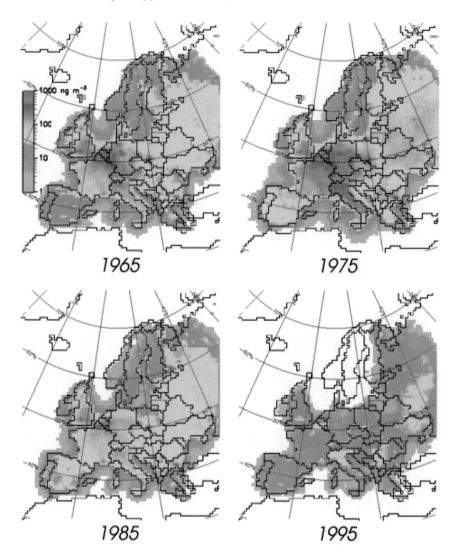

Fig. 5.30. Simulated annual mean lead concentrations in the planetary boundary layer [175]

to assess how much lead emitted in one country ends up being deposited in another country. The results indicate that most depositions in a country originate from its own emissions. Only smaller countries like Switzerland or the Netherlands have suffered substantial depositions from neighboring states: 20% of the Swiss depositions come from France, and 21% of the Dutch depositions from Belgium. As a further example the model data show that 23% of the total depositions over the Baltic Sea originate from Poland, 20% from

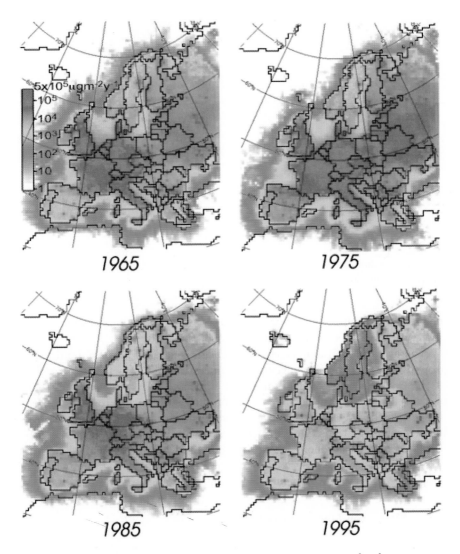

Fig. 5.31. Simulated annual mean lead depositions [175]

Germany, 16% from Finland, 12% from Sweden, 9% from the Russian Federation, 5% from Denmark and 1% from Norway. All other countries contribute less than 1%. Figure 5.32 shows the temporal evolution of the depositions. The simulated input into the Baltic Sea peaked in the mid 1970s, with 3500 tons annually, and has since then declined to less than 500 tons in 1995.

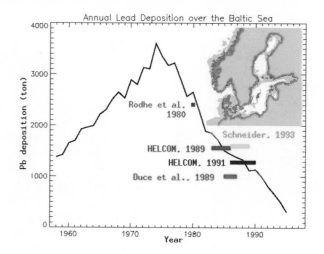

Fig. 5.32. Simulated input of lead into the Baltic Sea (*line*) and estimates based on comprehensive analyses of observational data [175]

The simulated data compare well with a comprehensive analysis based upon observational evidence in the second half of the 1980s[20].

The atmosphere (and the environment in general) will remain a dumping ground for various anthropogenic substances in the foreseeable future. Some of these substances will have negative impacts so that society will sooner or later begin to regulate the release of such substances. To do so, science has to provide society with tools to assess the situation in the past and the impacts of regulations. One such tool is a suitable model. Together with a detailed emission chronology and a regionalized history of weather events it can provide a detailed description of what has happened so far and what may happen as a consequence of future developments and regulations.

5.2.4 Altimeter Data and the Tides

Altimeter data from satellite missions are used for studies of the global ocean circulation and marine geophysics. A prominent example is the Topex/Poseidon mission. Its main purpose is to estimate the geostrophic surface currents from the low-frequency, non-tidal sea surface height field. The tides need to be removed for this purpose. However, the removal of the ocean tides, which are aliased in the raw data, is far from trivial. The tides are most accurately

---

[20] The effort to run the transport and deposition model was small compared to the effort to collect and analyze the measured data from the 1980s. The observational data are of course very important and needed to ensure the quality of the modeling effort. Models cannot replace monitoring efforts. But monitoring usually does not provide spatially and temporally complete descriptions. This requires models. Thus, the more successful monitoring systems combine observations and models.

determined by methods that combine tide gauge and altimeter data with a hydrodynamical tidal model. In this section we discuss such a data assimilation approach based on a paper by Egbert et al. [30].

1. Hydrodynamical calculation

The hydrodynamical *model* is based on Laplace tidal equations. A zeroth order version of them is given by the momentum balance (4.12), the volume balance (4.8) and the boundary conditions (4.4) (see Sects. 4.1.2–4.1.3). The equations require the specification of the tidal potential, either calculated as a sum over the contributions from different tidal frequencies or calculated from the actual celestial positions of the moon and sun. Often forcing at only a single tidal frequency is considered. Laplace tidal equations are solved numerically, usually on a spatial grid and by time stepping. The result of such a hydrodynamic calculation of the tides is given in Fig. 5.33 for the M_2 tide, the semi-diurnal tide with a period of 12 h and 25 min. The model integrates the full nonlinear Laplace tidal equations on a C-grid, forced at the M_2 tidal frequency. The resulting time series at each grid point are harmonically analysed to give the amplitude and phase shown in the figure.

When data are assimilated into such a tidal model one has to assign *error statistics* to the dynamical model. This requires considerable physical insight into the problem. There are first the errors associated with Laplace tidal equations. The volume balance equation (4.8) is a consequence of mass conservation for a Boussinesq fluid, which is an excellent approximation for the ocean. If discretization schemes are used that conserve volume then there is usually no or little error assigned to the volume balance. The same applies to the boundary condition (4.4) that there be no transports across coastlines. The situation is quite different for the momentum balance (4.12). Its form requires validity not only of the Boussinesq approximation but also of the shallow water approximation. It neglects the joint effect of baroclinicity and bottom relief that converts barotropic into baroclinic motion at sloping bottoms in a stratified ocean. The momentum balance also requires knowledge of the bottom topography H_0, which is a problem in some parts of the world's oceans. It requires the astronomical forcing ξ_{equ}, which is known extremely accurately though this accuracy is degraded once fudge factors like α and β are introduced to account for earth and load tides and self-attraction, as in (4.12). Another large error is introduced by the friction term $\mathbf{F}_h$ which represents unresolved subgridscale processes. Additional errors are of course introduced once the equation is discretized in space and time.

To assign a covariance matrix to all these errors does not only require estimates of error bars but also of correlations in space and among variables. There is no objective way to construct these covariance matrices. It is done subjectively by the user taking into account and trying to balance physical aspects, computational requirements and the purpose of the assimilation.

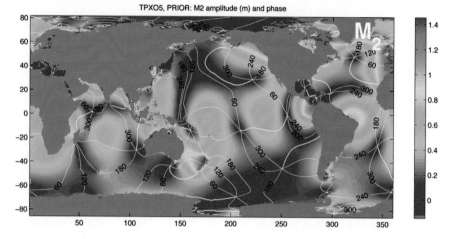

Fig. 5.33. Amplitude and phase of M$_2$-tidal elevation as simulated by a hydrodynamical tidal model without data assimilation. The elevation field was calculated by time-stepping the full nonlinear Laplace tidal equations on a C-grid with periodic forcing for 60 days followed by harmonic analysis of the last 40 days. Spatial resolution was 1/4 degree. Courtesy of Gary Egbert

2. Empirical calculations

Next we discuss the *data* that go into the assimilation scheme. Until the advent of satellite altimetry, data for tidal modeling came primarily from coastal, pelagic (open sea) and island tide gauges. Time series from such tide gauges were typically analyzed by assuming them to be of the harmonic form

$$\xi(t) = Re[\sum_{j=1}^{J} \xi_j exp(i\omega_j(t - t_0) + V_j(t_0))] + \delta\xi \tag{5.5}$$

Here $\omega_j, j = 1, \ldots, J$, are the tidal frequencies, ξ_j the complex tidal amplitudes, $V_j(t_0)$ the "astronomical" argument[21] at time t_0 for tidal harmonic j, and $\delta\xi$ an error term that accounts for non-tidal contributions to the surface elevation and other inaccuracies of the representation (5.5). The tidal coefficients ξ_j are then determined from long-term tide gauge data by harmonic analysis, assuming that the non-tidal contributions do not have any energy at tidal frequencies. The advantage of tide gauge records is that they usually have a sufficiently short sampling rate and long duration to allow for an accurate determination of the tidal amplitudes by harmonic analysis. Their main disadvantage is their spatial sparseness, especially for the determination of open ocean tides.

---

[21] The astronomical argument is the phase of the tidal potential for constituent j at Greenwich at time t_0. The phases of ξ_j are thus "Greenwich" phases.

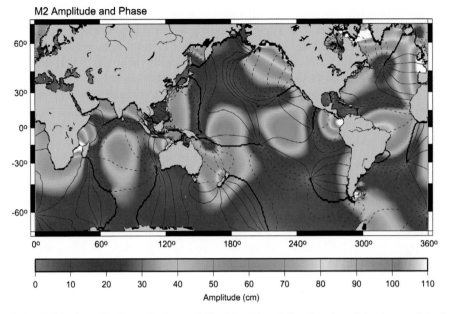

Fig. 5.34. Amplitude and phase of the M_2 tide of the Desai and Wahr empirical tidal model. The model is based solely on Topex/Poseidon altimeter data, on repeat cycles 10 to 356 which cover approximately 9.4 years. The contour interval for the phase is 30 degrees and dashed contours are negative phases [27] (© (1995) American Geophysical Union. Reproduced by permission of American Geophysical Union)

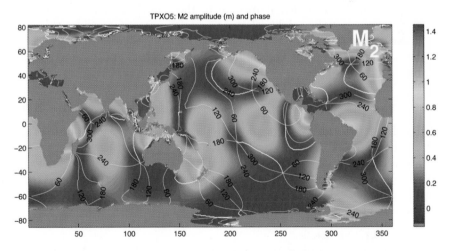

Fig. 5.35. Amplitude and phase of M_2 tide as obtained by assimilating satellite altimeter data into a hydrodynamical model. Courtesy of Gary Egbert

Satellite altimeter data now fill this gap with broad, accurate open-ocean coverage. In particular, the Topex/Poseidon satellite has been configured to accurately extract the tidal signal. A satellite altimeter measures the distance between the satellite and the sea surface. The satellite orbit can accurately be inferred from ground-based tracking stations, relative to a reference surface. The altimeter measurement thus gives the sea surface height relative to the same reference surface. Since distance is measured by the travel time of a radar impulse one needs to apply corrections to account for the delay in the atmosphere and for the effect of the sea state, but these are fairly well known. The major errors and complications come from three other sources:

1. The altimetric sea surface height gives the geocentric elevation $\xi_g = \xi + \xi_e + \xi_l$ instead of the water depth $\xi + H_0$[22]. The bottom elevation ξ_e of the earth tide is fairly well known and can be subtracted out. One thus usually considers the signal $\xi' = \xi_g - \xi_e$ in altimeter data processing. Assuming again that tidal loading, which is caused by the weight of the overlying water column, is just proportional to the tidal elevation one arrives at

$$\xi' = \tilde{\beta}\xi \tag{5.6}$$

with an appropriate constant $\tilde{\beta}$ that may be assumed to depend on the tidal constituent[23]. A simple estimate for the ocean tide signal is thus given by $\xi = \xi'/\tilde{\beta}$. More refined estimates can be obtained by considering different spherical harmonics of ξ separately and by performing the convolution with the appropriate elastic earth Green's function.

2. It is the sea surface height relative to the geoid[24] that is of dynamical significance, not the height relative to a reference surface. Since the geoid is not sufficiently well known at this time one eliminates its inaccuracies by considering the difference of measurements taken at the same position but at different times

$$d(\mathbf{x}, t_i) = \xi'(\mathbf{x}, t_i + \Delta t) - \xi'(\mathbf{x}, t_i) \tag{5.7}$$

For Topex/Poseidon data $\mathbf{x}$ would be the position of the ground track and Δt the period of the exact repeat cycle, or $\mathbf{x}$ would be the positions where ascending and descending ground tracks cross and Δt the period between such crossings, which depends on the position $\mathbf{x}$.

3. The complex tidal coefficients $\xi'_j = \tilde{\beta}\xi_j$ of the ξ' signal are estimated from the satellite data by fitting the Fourier expansion

---

[22] See Sect. 4.1.3 for details.

[23] Note that the parameter $\tilde{\beta}$ differs from the factor β that enters the momentum balance (4.12) of Sect. 4.1.3 and also contains the effect of gravitational self-attraction [137].

[24] The geoid is the surface of constant gravitational potential; its geocentric distance from the earth center is not uniform, but is distorted by the nonhomogeneous mass distributions in the solid earth.

$$d(\mathbf{x}, t_i) = Re[\sum_{j=1}^{J} \tilde{\beta}\xi_j(\mathbf{x})(e^{i\phi_j(t_i+\Delta t)} - e^{i\phi_j(t_i)})] + \delta d \qquad (5.8)$$

to the difference data. Here, the main problem is caused by the shortness and poor sampling of this difference time series that makes it hard to extract all relevant tidal amplitudes by harmonic analysis. Additional assumptions have to be invoked to do so at least approximately.

The determination of the tides from measurement alone is called an empirical tidal model. The only empirical model for Topex/Poseidon data that we are aware of is the one of Desai and Wahr [27]. Its results for the M_2-tide are reproduced here in Fig. 5.34. It shows the same broad features as the hydrodynamical model in Fig. 5.33 and differences are hard to detect, but taking differences between the empirical and hydrodynamic model reveals differences with amplitudes of up to 10 cm over large parts of the open ocean, and more in shallow seas. Of course, the empirical model only gives the surface elevation and not the currents or the transports, which can also be calculated from the hydrodynamical model.

From a *data assimilation* point of view, (5.8) is the measurement model. The state vector is composed of the tidal coefficients $\xi_j(\mathbf{x})$ and $\mathbf{U}_{h,j}(\mathbf{x})$, and $d(\mathbf{x}, t_i)$ are the data. Again considerable insight must be exercised when assigning errors δd to this measurement equation. These errors include not only the inaccuracies of the altimeter data but also the strength of the non-tidal signals.

3. Data assimilation

The combination of data and model in [30] is done by means of an inverse method (see Appendix D). The result of the inversion is given in Fig. 5.35 for the M_2 tidal amplitude and phase. By carefully comparing this data assimilation estimate with the hydrodynamic estimate (Fig. 5.33) and the empirical estimate (Fig. 5.34) one can assess where and how assimilation has improved the tidal estimate. Overall one finds the following gains:

- The assimilation solution provides more realistic estimates of the tidal signal in shallow seas where altimeter data are unable to resolve the generally small spatial scales.
- While the assimilation does not much reduce the uncertainties in the elevation field it provides a much smoother solution. Thus elevation *gradients* have significantly smaller errors. This has implications for the residual tidal errors in geostrophic current estimates.
- Assimilation provides the tidal currents. These currents are required for the dissipation calculation that we discuss in Sect. 6.3.3, and are useful for other purposes as well, such as correcting vessel-mounted Acoustic Doppler Current Profilers (ADCP), providing barotropic tidal boundary conditions for regional models and for calculation of angular momentum budgets.

- Assimilation allows one to quantify the errors or residuals in the dynamical equations. This may lead to the identification of missing physics, as shown in Sect. 6.3.3, where *deep ocean* tidal dissipation suggests itself as a major contributor to the tidal energy balance.

5.3 Scenarios

The art of building scenarios is technically similar to the art of forecasting. Forecasts assume that the information provided by initial and boundary conditions and by parameters is real. Scenarios anticipate the results of plausible, not necessarily likely changes in these conditions and parameters. Examples are climate changes in response to increasing atmospheric concentrations of greenhouse gases, changes in the tidal regime due to the dredging of a tidal channel, the effect of a new regulatory system on the water quality of a river, or changes in an ecosystem exposed to changing concentrations of nutrients.

When released to the public, or a wider scientific community, scenarios often undergo a metamorphosis from the originally intended *storyline* of something plausible, or possible, but not necessarily probable, to an almost certain forecast. This is the more frequent, the more practical implications a scenario has. An example is the anthropogenic climate change due to greenhouse gases. When this process was first studied, it was usually done within the *doubling CO_2 format*. Two extended equilibrium simulations with an atmospheric general circulation model (GCM) coupled to an oceanic GCM or surface mixed layer model were made, one with the present greenhouse gas levels, and another one with levels corresponding to a doubling of the pre-industrial level. This format was used because computer resources at that time did not allow the continuous integration of climate models that included a gradual increase of greenhouse gas concentrations from present or pre-industrial levels to increased levels anticipated for, say, the end of the 21st century. The reliance on the "doubling CO_2" format led to the widespread perception that climate change means that the climate changes from one equilibrium state to another, from "$1 \times CO_2$" to "$2 \times CO_2$". The water level along a coast would rise from one value to another value. The essential aspect that climate will not be in equilibrium for an extended time because of the ongoing emissions of greenhouse gases is completely lost in this view. Of course, such an interpretation is entirely inadequate, but perhaps unavoidable given the dynamics of public attention and media presentation.

Scenarios are meant to guide experts as well as the public at large so that options for decisions and actions can be weighted with the associated risks. Scenarios are brainstorming tools to answer the question *what happens if*. They are particularly valuable when the vulnerability of the environment and options for adaptation measures are explored[25]. Almost always several

---

[25] Schwartz [147] offers an interesting overview about the use of scenarios in preparing business decisions.

alternative scenarios are needed, to contrast different future developments. Scenarios are not the most probable future developments.

In the following we limit ourselves to one example, the widely discussed and politically charged climate change scenarios prepared by the Intergovernmental Panel on Climate Change (IPCC).

Climate change scenarios are conditional forecasts, forecasts of the second kind. They utilize scenarios of greenhouse gas or aerosol emissions, or of changing land use, to estimate how the climate statistics may change. They are conditioned on these emissions or changes in land-use. The scenarios of changing emissions and land use are prepared by economists and other social scientists. Examples are the four "SRES"[26] scenarios prepared for the "Third Assessment Report" of the IPCC:

(A1)   a world of rapid economic growth and rapid introduction of new and more efficient technology[27],

(A2)   a very heterogeneous world with an emphasis on family values and local traditions[28],

---

[26] "IPCC Special Report on Emissions Scenarios", see http://www.grida.no/climate/ipcc/emission/. The scenarios do not anticipate any specific mitigation policies for avoiding climate change. The authors emphasize that "no explicit judgments have been made by the SRES team as to their desirability or probability".

[27] This scenario is described by SRES as follows: "A case of rapid and successful economic development, in which regional average income per capita converge – current distinctions between "poor" and "rich" countries eventually dissolve. The primary dynamics are: Strong commitment to market-based solutions. High savings and commitment to education at the household level. High rates of investment and innovation in education, technology, and institutions at the national and international level. International mobility of people, ideas, and technology. The transition to economic convergence results from advances in transport and communication technology, shifts in national policies on immigration and education, and international cooperation in the development of national and international institutions that enhance productivity growth and technology diffusion. The global economy expands at an average annual rate of about 3% to 2100. Energy and mineral resources are abundant because of rapid technical progress, which both reduces the resources needed to produce a given level of output and increases the economically recoverable reserves. Energy use per unit of GDP decreases at an average annual rate of 1.3%. The concept of environmental quality changes from the current emphasis on "conservation" of nature to active "management" of natural and environmental services. With the rapid increase in income, dietary patterns shift initially toward increased consumption of meat and dairy products, but may decrease subsequently with increasing emphasis on the health of an aging society. High incomes also translate into high car ownership, sprawling suburbia, and dense transport networks, nationally and internationally." (abbreviated version)

[28] "...characterized by lower trade flows, relatively slow capital stock turnover, and slower technological change. The world "consolidates" into a series of economic regions. Self-reliance in terms of resources and less emphasis on economic, so-

(B1)   a world of "dematerialization" and introduction of clean technologies[29], and

(B2)   a world with an emphasis on local solutions to economic and environmental sustainability[30].

---

cial, and cultural interactions between regions are characteristic for this future. Economic growth is uneven and the income gap between now-industrialized and developing parts of the world does not narrow. People, ideas, and capital are less mobile so that technology diffuses more slowly. International disparities in productivity, and hence income per capita, are largely maintained or increased in absolute terms. With the emphasis on family and community life, fertility rates decline relatively slowly, which makes the population the largest among the storylines (15 billion by 2100). Technological change is more heterogeneous. Regions with abundant energy and mineral resources evolve more resource-intensive economies, while those poor in resources place a very high priority on minimizing import dependence through technological innovation to improve resource efficiency and make use of substitute inputs. Energy use per unit of GDP declines with a pace of 0.5 to 0.7% per year. Social and political structures diversify; some regions move toward stronger welfare systems and reduced income inequality, while others move toward "leaner" government and more heterogeneous income distributions. With substantial food requirements, agricultural productivity is one of the main focus areas for innovation and research, development efforts, and environmental concerns. Global environmental concerns are relatively weak." (abbreviated version)

[29] "The central elements are a high level of environmental and social consciousness combined with a globally coherent approach to a more sustainable development. Governments, businesses, the media, and the public pay increased attention to the environmental and social aspects of development. Economic development is balanced, and efforts to achieve equitable income distribution are effective. This is a fast-changing and convergent world which invests a large part of its gains in improved efficiency of resource use ("dematerialization"), equity, social institutions, and environmental protection. Particular effort is devoted to increases in resource efficiency to achieve the goals stated above. Organizational measures are adopted to reduce material wastage by maximizing reuse and recycling. Global population reaches nine billion by 2050 and declines to about seven billion by 2100. This is a world with high levels of economic activity and significant and deliberate progress toward international and national income equality. A higher proportion of this income is spent on services rather than on material goods, and on quality rather than quantity. A strong welfare net prevents social exclusion on the basis of poverty." (abbreviated version)

[30] "...increased concern for environmental and social sustainability. Increasingly, government policies and business strategies at the national and local levels are influenced by environmentally aware citizens, with a trend toward local self-reliance and stronger communities. Human welfare, equality, and environmental protection all have high priority, and they are addressed through community-based social solutions in addition to technical solutions. Education and welfare programs are pursued widely, which reduces mortality and fertility. The population reaches about 10 billion people by 2100. Income per capita grows at an intermediate rate. The high educational levels promote both development and environmental

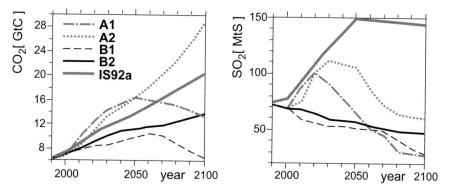

Fig. 5.36. SRES scenarios of carbon dioxide (*left*) and sulfate emissions (*right*) constructed for the Third Assessment Report of the IPCC. The scenarios A1, A2, B1 and B2 are explained in the text, the scenario IS92a was constructed for the IPCC Second Assessment Report in 1995

These SRES scenarios translate into the expected emissions of greenhouse gases and sulfate aerosols, as given in Fig. 5.36. These emissions are fed into climate models. It is again stressed that the output of climate models will depend on these emissions and hence on the *assumptions* about future social, economic and technical developments.

For two of the scenarios, A2 and B2, the temperature change simulated by the climate model of the Danish Meteorological Institute, averaged over the years 2070 to 2100, is displayed in Fig. 5.37 [158]. Almost everywhere a temperature increase is simulated, except for isolated areas in the Southern Ocean. The warming is most pronounced over land and over the Arctic, where sea ice is expected to partially vanish. Temperatures of sea ice can be very low, while the minimum temperature of open water is about $-1°C$. Accordingly, the air temperature above an area previously covered by sea ice will increase

protection. Environmental protection is one of the few truly international common priorities. However, strategies to address global environmental challenges are not of a central priority and are thus less successful compared to local and regional environmental response strategies. The governments have difficulty designing and implementing agreements that combine global environmental protection. Land-use management becomes better integrated at the local level. Urban and transport infrastructure is a particular focus of community innovation, and contributes to a low level of car dependence and less urban sprawl. An emphasis on food self-reliance contributes to a shift in dietary patterns toward local products, with relatively low meat consumption in countries with high population densities. Energy systems differ from region to region. The need to use energy and other resources more efficiently spurs the development of less carbon-intensive technology in some regions. Although globally the energy system remains predominantly hydrocarbon-based, a gradual transition occurs away from the current share of fossil resources in world energy supply." (abbreviated version)

quite substantially. By and large, the difference between the two scenarios is not very large.

Such a temperature increase is a common result of all models used to construct climate change scenarios. Also, the models agree that the thermal expansion of sea water will lead to an increase of sea level, which may be dampened or compounded by the growing or melting of the two big ice sheets, Antarctica and Greenland. The fact that the models agree lends some confidence to these results, but is no proof that the real world would actually develop as the models predict, even if the emissions were exactly as prescribed. The development of climate models and their sensitivity to human interference is so closely scrutinized by the scientific community that dramatic deviations from contemporary perceptions about the functioning of the climate system and of its sensitivity are not well received and usually quickly rectified. Thus, the development of climate models underlies to some extent a social control.

There is also some convergence among the models on precipitation statistics, while other aspects, like storm activity, are reproduced less consistently by the different models, and the Intergovernmental Panel on Climate Change takes a cautious stand on making statements about these matters.

The temperature maps shown in Fig. 5.37 seem to provide an assessment on a very detailed spatial scale. This impression is, unfortunately, incorrect. The problem with Fig. 5.37 is that it combines two types of graphical presentations, namely the continuous coast line of the world and the discrete matrix of about 2000 grid points, given by the color code. Thus, even if the map contains the Hawaiian Islands, the model does not contain them. The immediate response of lay people is to look at what is expected to happen at "their" location. This is futile as the models are not capable of simulating local specifics. Thus, the presentation adopted in Fig. 5.37 might be quite misleading for non-experts, even though it is commonly used in presenting climate change scenarios.

In an effort to sort out to what extent regional assessments of climate change are robust, an IPCC working group compared expected regional mean changes in a variety of simulations [47]. For a series of sub-continental regions ($> 10^7 \, \mathrm{km}^2$), they determined whether at least 7 out of the 9 models used in the Third Assessment Report, and 4 out of the 5 models used in the previous Second Assessment Report arrived at consistent changes in terms of precipitation and air temperature. The Second Report [65] used somewhat different scenarios, in particular one, labelled GG, with a "business-as-usual" increase in greenhouse gas emissions and a second, labelled GS, where aerosols are released into the atmosphere in addition to greenhouse gases. Note also that the models used in the Second and Third Reports differ, as they all are continuously improved and updated.

Figures 5.38 and 5.39 represent the result of the effort. Interestingly, both the temperature and precipitation changes seem to be consistent in most areas, both in winter and summer, not only among the models used in the Second or Third Report, but also across the different scenarios A2, B2, GG and GS.

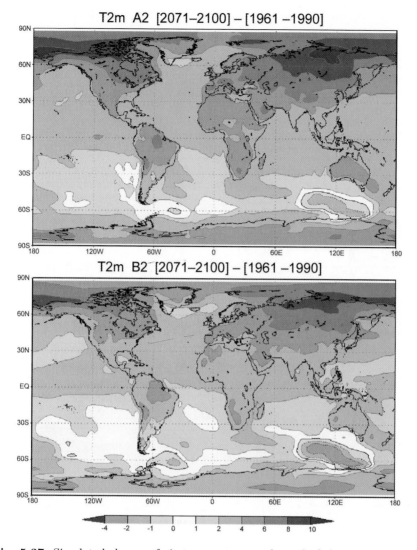

Fig. 5.37. Simulated change of air temperature at the end of the 21st century in two climate change simulations forced with emission scenarios A2 and B2. Courtesy: Danmarks Meteorologisk Institut

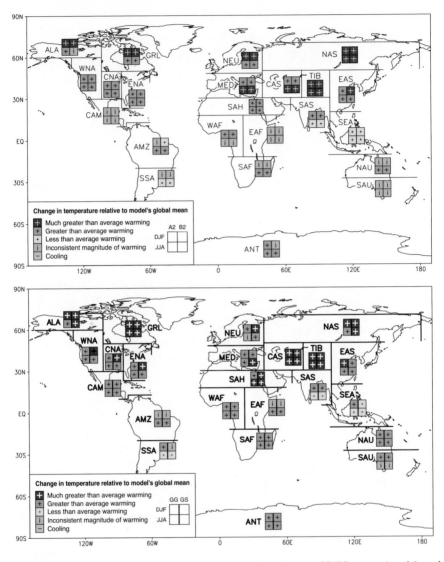

Fig. 5.38. Regionalized change in temperature for the two SRES scenarios A2 and B2 (*top*), and for two business-as-usual emission scenarios of the Second Assessment Report, with and without aerosol emissions (GG and GS) (*bottom*). For each region, a box with four entries is given, describing the stability of the simulated change in the two scenarios (*columns*) for two seasons (northern winter DJF, and summer, JJA; *rows*). If 7 out of 9 (*top*) or 4 out of 5 (*bottom*) models agree on a change, then a "+" (warming) or a "–" (cooling) is inserted in the box. If the models do not agree, then an "i" is given [47]

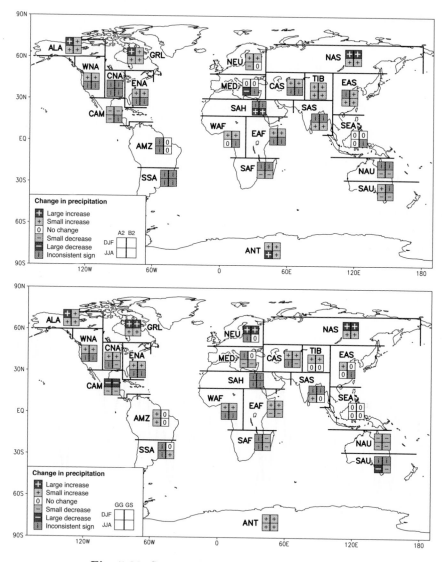

Fig. 5.39. Same as Fig. 5.38 but for precipitation

This convergence adds to the credibility of the models, but it is no guarantee that the models might fail to describe certain, so far not identified significant processes. Convergence among different models is no proof of their realism.

5.4 Secondary Applications

The examples given so far in this chapter cover the most important applications of models in atmospheric and oceanic sciences. In addition there are a number of secondary applications, among them are:

The design of efficient observational networks

In many cases there are a limited number of "hot spots", the knowledge of which is representative for the state of the system as a whole. A model simulation can help to identify these. An example is Janssen's [70] simulation of the state of the Baltic Sea. In this simulation, the tide gauge of Landsoort in Sweden (close to Stockholm) emerged as being representative for the water content in the Baltic Sea (Fig. 5.40). This model result is supported by empirical evidence.

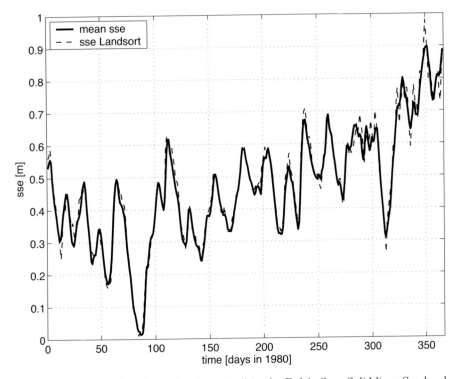

Fig. 5.40. Time series of simulated sea level in the Baltic Sea. *Solid line:* Sea level elevation at Landsoort south of Stockholm; *dotted:* Area average of sea level in the Baltic Sea. From Janssen [70]

Estimation of rare events

In many cases, the observational evidence is not sufficient to estimate the tails of the distributions of relevant variables. Examples are extreme values of ocean wave heights or the natural range of climate variability. With long simulations such statistics can easily be obtained. Of course, estimates based on such simulations are *neutral analogs*, and it remains to be demonstrated that the models provide realistic estimates.

Climate change detection and attribution

An important task of climate science is the examination of the ongoing climate variability. The problem is whether it is in the range of natural variations (detection) or, if not, whether it can plausibly be explained by forcing factors such as elevated greenhouse gas concentrations (attribution) [59]. For both, the detection and attribution problem, the output of models is used to guide the analysis.

The global detection problem is a multivariate hypothesis test with a large number of degrees of freedom. Without prior knowledge, chances to formulate a-priori the "right" null hypothesis are slim, and thus the chances to detect anthropogenic climate change are also slim [56]. However, with the help of a quasi-realistic climate model, a likely pattern of climate change can be obtained and used as a *first guess* in the detection exercise and increases the chances to detect changes. For a further review of the detection problem refer to [197].

Similarly, the attribution problem is analyzed by assessing how well the response of a quasi-realistic climate model to a variety of potential forcing factors explains the observed climate change [55].

6

Modeling in Fundamental Environmental Sciences – Simulation and Hypothesis Testing

In the previous chapter we have presented examples of the utility of quasi-realistic models in forecasting the future state of a system, in describing the state of a system, and in studying plausible scenarios. These efforts generally result in large data sets, which then serve as input for other studies, such as assessments of past systematic changes, the estimation of the intensity of rare events, or decisions about future activities. In this chapter a different type of utility of quasi-realistic models is demonstrated with a series of examples, namely their role in answering specific scientific questions. Admittedly, the borderline between Chaps. 5 and 6 is not well-defined, but we think that our discrimination clarifies an important issue, namely that quasi-realistic models play a dual role, as generator of a substitute reality for all kinds of applied studies and as a tool to investigate fundamental scientific questions. This chapter is about the latter and the following examples will illustrate

- the option to test the validity of hypotheses, derived from theoretical considerations, empirical evidence or pure speculation (Sect. 6.1),
- the use of quasi-realistic model output (in the spirit of Sects. 5.2 and 5.3) to infer parameters of reduced models and to determine the skill of such models in conceptualizing aspects or parts of reality (Sect. 6.2), and
- the possibility to infer characteristics of the system under consideration, which cannot be derived from observations (Sect. 6.3).

6.1 Hypothesis Testing

In this section, we present examples where quasi-realistic models are used to test specific hypotheses brought forward by theoretical reasoning, plausibility arguments or mere speculation. The quasi-realistic models serve as a *substitute reality* or *virtual laboratory*. One follows the methodological approach of classical natural sciences by performing experiments. The hypothesis may be falsified by the experiment but not verified. The experiment is, however, not a real-world analog (i.e., a *laboratory model*) of a certain idealized situation, but a numerical analog (i.e., a *numerical model*). In both cases inconsistency of the experiment's outcome with reality may signal the failure of the hypothesis or the failure of the analog construction. The models may simply be inadequate to describe the situation addressed.

The following cases are presented:

- The first case, drawn from [163], addresses the question of whether tidal currents in a small tidal inlet, with dimensions of a few kilometers, are significantly affected by the Coriolis force. Back in the early 1970s this question was far from being a mere academic problem. At that time quasi-realistic modeling of tides in such inlets was done with hydraulic models (see Sect. 1.3.2), which were incapable of accounting for the Coriolis effect. The numerical experiment would determine their fate.

- Past climates can be hypothesized as being essentially the response to anomalous forcing conditions, such as solar output or the presence of stratospheric aerosols. Such a hypothesis can be tested by comparing the output of a quasi-realistic climate model with historical reconstructions of climate from indirect or proxy data, such as tree ring characteristics or depositions in ice cores. Our example deals with the *Late Maunder Minimum* of the sunspot cycle from 1675 to 1710.

- Originally, most people believed that all climate variability was caused by external forces, but in 1976 Hasselmann [58] introduced the *stochastic climate model* which ascribes climate variability to internal sources, namely random weather fluctuations. Our example, drawn from [115], demonstrates that a quasi-realistic ocean model exposed to random forcing indeed produces variations as predicted by the stochastic climate model.

- The stability of the *thermohaline circulation* in the North Atlantic is another open problem of current climate science. In one of his many seminal papers, Stommel [161] offered a conceptual model. It describes the efficiency, or shut-down, of this overturning circulation in response to the temperature and salt fluxes at the surface of the ocean on time scales of hundreds or thousands of years. Obviously, this model cannot be proven experimentally. Observational evidence from paleoclimatic proxy data can at best serve as a rough check. A much more comprehensive test of Stommel's model can be achieved with quasi-realistic ocean models. Our example is that of Rahmstorf [136].

- A variety of mechanisms have been suggested to explain Alpine lee cyclogenesis. A strategy to test these theories was devised by Egger [33] and Tafferner and Egger [165]. The key idea is to first find a quasi-realistic model that is capable of realistically simulating the process of Alpine lee cyclogenesis. Then, as a second step, the model is manipulated so that the suggested mechanisms control the model's dynamics. The crucial test is then whether or not Alpine lee cyclogenesis also takes place in the manipulated model.

6.1.1 Tides and the Coriolis Force

Here we revisit the case of the tidal flow in the tidal inlet Jade Bay in the southern German Bight. Two models have been presented for this problem, a hydraulic model (Fig. 1.4) and a mathematical model, based on Laplace tidal equations (Sect. 4.1.4).

The first research question is whether the two models, the hydraulic and the mathematical one, are consistent[1]. If they are, then the costly hydraulic models may in many cases be replaced by much more economic mathematical models. Thus, the interest is not so much in the details of the tidal currents in Jade Bay but in the more fundamental problem of whether numerical models are adequate tools for simulating tidal phenomena.

The question "are the models consistent?" is not really well posed. An all encompassing comparison cannot be carried out. Instead, the models are considered consistent as long as no *major* differences in their performance are found. Consistency between models as well as between model and reality can be checked by very different means. Here, we limit ourselves to the simplest strategy, namely to the comparison of states simulated by the two models.

Figure 1.6 displayed the circulation of the hydraulic model and of the numerical model *without* Coriolis force shortly after high tide. The main features of both circulation patterns are similar. The mathematical and hydraulic model return consistent velocities. We may be tempted to answer the first research question "are the models consistent" positively.

As a second research question we ask a typical fundamental research question, namely what is the relative importance of a certain process. The mathematical and the hydraulic model have been run without the Coriolis force. The question is whether the Coriolis force is truly of secondary importance for the tidal currents. In the framework of a mathematical model, this problem is easily addressed in a *numerical experiment*. One simulation is done without the Coriolis force, and another one with the Coriolis force. If the two model versions return similar results, then the Coriolis force may be considered unimportant.

These two experiments have been carried out by Sündermann and Vollmers [163]. It turns out that the difference is not small but large. As can be seen in Fig. 6.1, the circulation with Coriolis force is no longer symmetric, as in Fig. 1.6, but strongly asymmetric. This finding implies that the Coriolis force *is* of major importance. The mathematical model does not only match the skill of the hydraulic model but actually surpasses it since it can include the Coriolis force and provide more realistic simulations.

In fact, the earlier wide-spread and well developed use of hydraulic and other mechanical models has disappeared, except for the simulation of very small scale processes, such as the dispersion of chemical pollutants in accidents (cf., [145]).

---

[1] One may argue, whether this question is a case of applied or fundamental research.

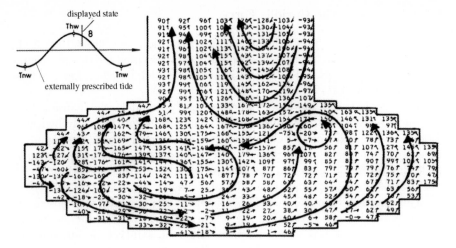

Fig. 6.1. Tidal currents in the numerical model. The timing is given by the inset: the tide has just passed the peak level and the water begins to flow out of the basin. This figure should be compared with Fig. 1.6. From Sündermann and Vollmers [163]

6.1.2 The Sun and the Late Maunder Minimum

The *Late Maunder Minimum* (LMM; AD 1675 to 1710) refers to a period with a particularly low number of sunspots and reduced solar activity. This period coincides with several large volcanic eruptions. Historical reports and proxy data document a substantial cooling during the winter half year at the end of the 17th century in Europe. It has been proposed that the anomalous solar output and, possibly, the simultaneous volcanic activity, may be the cause for this observed cooling. It has further been suggested that the cooling was not limited to Europe, but a more widespread if not global event involving anomalous conditions in the North Atlantic [92].

In order to test these hypotheses, a multi-century integration was carried out by Zorita et al. [196] with the climate model ECHO–G, which is a combination of the ocean model HOPE–G in T42 resolution and the atmospheric model ECHAM4 in T30 resolution. The simulation extended over 250 years from 1550 to 1800. The model was driven by the time-varying solar constant shown in Fig. 6.2, which represents the solar and volcanic activity and the changing atmospheric concentrations of greenhouse gases during that period[2]. The spikes represent the effect of the volcanic eruptions, while the

---

[2] The atmospheric CO_2 and methane concentrations were derived from air trapped in Antarctic ice cores. The variations of solar output and the influence of volcanic aerosols on the radiative forcing were derived from the number of sun spots after 1600 AD and concentrations of cosmogenic isotopes in the atmosphere before 1600 AD. The forcing due to volcanic aerosols was estimated from concentrations of

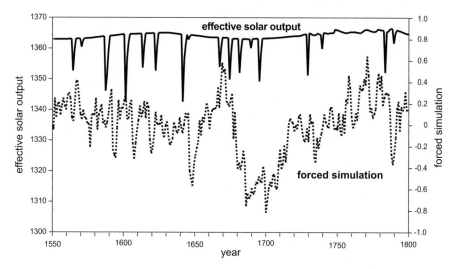

Fig. 6.2. Solar forcing and Northern Hemisphere mean temperature during 1550–1800 as simulated in the Late Maunder Minimum experiment

slow variation reflects the changing output of the sun. The greenhouse gas concentrations vary only insignificantly during this time.

The following strategy was applied: As the hypotheses refer to Northern winter conditions, the analysis was limited to the DJF season. First, it was shown that during the time of the LMM (1675–1710), the model generated a marked cooling in Europe. Next, it was shown that the simulated temperatures in Europe during this period are consistent with the limited observational evidence constructed from early instrumental time series and proxy data. The timing, pattern and intensity of the historical event are successfully reproduced by the model, as far as it has been reconstructed. The model is thus "validated". The first hypothesis, the causal link between solar and volcanic activity and the cooling in Europe, is considered to be consistent with the experiment (i.e., it is not falsified). Finally, the model was used in a constructive manner. It was assumed that the simulation is not only realistic in Europe but also globally. A persistent temperature reduction, for almost the entire globe north of 30°S is found. A particularly strong anomaly of temperature and sea ice concentration is formed in the Labrador Sea. A significant change in the North Atlantic overturning was not found. These properties are neutral analogs. The more detailed story follows.

At the end of the 17th century, beginning at about 1675 and ending in the early 18th century, the model develops a marked temperature anomaly, with an amplitude of more than 0.5 K, as can be seen in Fig. 6.2. This anomaly does not emerge in simulations where the radiative forcing is constant. It does

---

sulphuric compounds in various ice cores, located mainly over Greenland. These forcing factors were translated into effective variations of the solar constant.

develop in simulations with the time-varying forcing, even when the model is started from different initial conditions. Therefore, the event is probably related to the anomalous radiative conditions – supporting the first hypothesis.

The spatial distribution of the simulated winter temperature difference between 1675–1710 and 1550–1800 is displayed for Europe[3] in Fig 6.3, together with a significance measure. The model generates a cooling almost everywhere, with maximum values of 1.2 K and more in Eastern Europe and northwest Russia/northern Norway, and of 0.5 K and more over most of Europe. No regions of warming are simulated. To exclude the possibility that the differences shown in Fig. 6.3 are the mere results of random variations, a series of local t-tests was conducted. The "control" period, 1550–1675 and 1710–1800, is compared with the "treatment period" 1675–1710. Areas are hatched in Fig. 6.3 whenever the null hypothesis of equal means is rejected. The different hatchings indicate different significance levels: dense hatching indicates a 99%, medium hatching a 95%, and minimum hatching a 90% significance level. The figure shows that the simulated cooling in Europe is almost everywhere robust, except for part of Northern Scandinavia.

Figure 6.4 shows the "observed" winter (DJF) temperature over Europe during the LMM, as reconstructed by Luterbacher et al. [106] from a combination of early instrumental time series and documentary evidence[4]. The reconstruction shows that cooling was widespread in Europe, with maximum values of more than 1.2 K in Eastern Europe and 0.5 K in many parts of Germany, France, Italy and Greece. A warming shows up only in a few locations, in northern Norway, Iceland and southeastern Greenland. There is also evidence of cooling in the Russian North, starting at the beginning of the LMM, but this cooling has not been quantified and is not included in the figure.

The hatching in the figure describes the proportion of explained variance when the reconstruction method is applied to data from the period 1961–1990. In areas with dense hatching, more than 90% of the variance is recovered, in areas with medium hatching it is more than 70% and in areas with light hatching it is more than 40% [105]. Accordingly, the confidence in the temperature

---

[3] The boxes clearly depict the coarse approximation of the European land mass by the T30 representation.

[4] Luterbacher et al. [106] used a large number of independent variables for their reconstruction. As predictors, they used instrumentally based variables such as the Central England temperature and the Paris Station pressure and temperature. Most of the predictors were, however, temperature and precipitation indices from sites all over Europe. These indices were estimated from high-resolution documentary evidence that included observations of ice and snow features, and other phenomenological and biological observations. A particular example is the ice cover of the western Baltic Sea (Fig. 4.10).

The predictand was the binned DJF mean temperature in Europe. Predictand and predictor were related through a linear regression. The performance of the regression model was tested by comparing the estimated temperatures with observed temperatures during the 20th century.

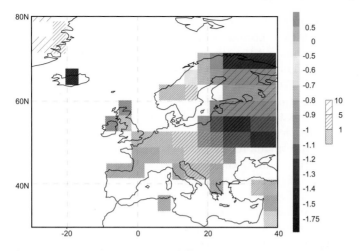

Fig. 6.3. Simulated winter temperature difference between the Late Maunder Minimum and the period before and after. Overlaid is the risk with which the local null hypothesis of equal time means is rejected. Dense hatching: $\leq 1\%$, medium: $\leq 5\%$ and light hatching: $\leq 10\%$

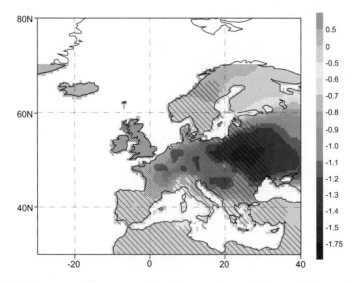

Fig. 6.4. Best guess of European winter temperature difference between the LMM and the periods before and after. The temperature is constructed with a regression model using a variety of local data in Europe. The proportion of explained variance, derived for the test period 1961–90, is indicated by the hatching. Dense: $\geq 90\%$, medium: $\geq 70\%$ and light: $\geq 40\%$

estimates is low in many areas, particularly in Finland and Russia and in the allegedly warmer-than-normal areas.

Little is known about the temperature conditions during the LMM outside Europe. Historical documentary records from China indicate a warming trend from a cold mid-17th century to a mild 18th century and a weakening of winter westerlies during the LMM. Coral records from the Galapagos, New Caledonia, the Great Barrier Reef and off Madagascar [194] indicate a general cooling in the tropics of about 0.5 K during the LMM. In the United States and southwestern Canada reconstructions from tree rings show distinct lower annual temperatures during the LMM. Especially in the eastern United States the last 30 years of the 17th century were the coldest for at least the last 400 years.

The model reproduces some of these global features. The model also contains climatic noise on all space and time scales. One can thus not expect that the model reproduces all details of the spatial patterns and temporal evolution. One can, however, expect that the model reproduces the *broad* spatial patterns as well as the *mean* evolution, which it does. Thus, the model result is considered a model realization of a "Late Maunder Minimum" cooling event, triggered by external solar and volcanic forcing.

Formally, the cooling over Europe in the model and the historical cooling over Europe represent *positive analogs* in the sense of Hesse (Sect. 3.1). In the next step the model is used constructively. It is assumed that other large scale aspects of the simulations are realistic as well; neutral analogs are assumed to be actually positive ones.

The global simulated temperature anomaly distribution during the LMM is displayed in Fig. 6.5. It shows an almost global cooling. The cooling is particularly pronounced, with 1 K and more, in northeast Canada, Greenland, the northern North Atlantic. The ice coverage of the Labrador Sea increased by up to 25% (not shown). The Northern Hemisphere continents show a weaker cooling, between 0.5 K and 1 K. Over most of the rest of the globe north of 30°S a cooling of up to 0.5 K is simulated, with isolated regions of warming in the Southern Hemisphere. Thus, the simulated event is a global phenomenon. These model results seem to be consistent with a variety of proxy-data [194].

6.1.3 The Stochastic Climate Model at Work

The cause of climate variability is obviously a relevant question. Originally, climate variability was believed to be caused by time-varying external forces, especially by orbital variations and by the solar cycle, but other non-cyclical "forces" such as deforestation were also thought to be a major cause of "systematic" climate change[5]. Then in 1976, Hasselmann [58] introduced the stochastic climate model. He asserted that most climate variations are forced

---

[5] An account of these ideas for the late 19th century is given in [155] and for the early 20th century in [67].

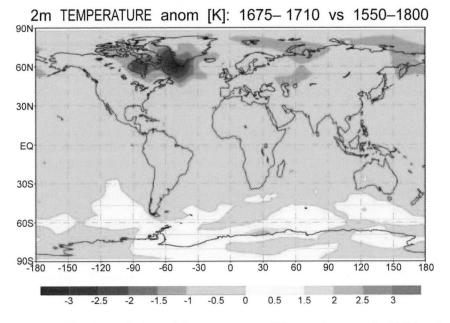

Fig. 6.5. Global distribution of the temperature difference between the LMM and the period before and after, as simulated by the climate model with anomalous radiative forcing

internally by the ever-changing unpredictable weather fluctuations, in much the same way that fluid molecules force Brownian ink particles. Weather fluctuations are generated by atmospheric chaotic dynamics. They manifest themselves, however, not as aesthetically pleasing patterns like the Lorenz attractor but as irregular fluctuations hardly distinguishable from *stochastic noise*. The slower components of the climate system, like the ocean or ice sheets "integrate" this mostly "white" noise, and eventually show marked long-term irregular variations. The simplest conceptual model for this stochastic climate model is an autoregressive model $\mathbf{X}_t$ of first or second order. The first order process is called *red noise*[6] and given by

$$\mathbf{X}_t = \alpha_1 \mathbf{X}_{t-1} + \mathbf{N}_t \qquad (6.1)$$

with a constant α_1 and a white noise term $\mathbf{N}_t$[7]. Without the noise term and $|\alpha_1| < 1$ equation (6.1) describes a damped system. Any initial pertur-

---

[6] See Appendix C.1.3. Red noise processes are also called AR(1) processes and second order processes AR(2).

[7] Here, white noise is a discrete stochastic process, indexed by integers $t = -\infty, \ldots -1, 0, +1, +2, \ldots \infty$. At each time t numbers are drawn from the *same* random distribution. Realizations at any two different times t_1 and t_2 are independent. For a more elaborate discussion, see Sect. C.1.3 and [182].

bation decays away within a characteristic time. The noise term randomly excites this damped system and causes irregular behavior including persistent anomalies. These low frequency variations are the climate variations excited by the random white noise weather fluctuations.

Second order autoregressive processes

$$\mathbf{X}_t = \alpha_1 \mathbf{X}_{t-1} + \alpha_2 \mathbf{X}_{t-2} + \mathbf{N}_t \tag{6.2}$$

with constants α_1 and α_2 and noise term $\mathbf{N}_t$ represent a slightly more complex stochastic climate model. Time series generated by this process also vary mostly irregularly, but for limited times, segments will exhibit certain characteristic features such as intermittent oscillatory behavior.

The continuous version of the autoregressive process (6.1) is the Langevin equation

$$\frac{d}{dt}\mathbf{X}(t) = -\alpha\mathbf{X}(t) + \mathbf{N}(t) \tag{6.3}$$

with a continuous white noise term $\mathbf{N}(t)$ that satisfies $E[\mathbf{N}(t)] = 0$ and $COV[\mathbf{N}(t), \mathbf{N}(t')] = \sigma^2\delta(t - t')$. The spectrum of the process $\mathbf{X}(t)$ is given by

$$S_\mathbf{X}(\omega) = \frac{\sigma^2}{2\pi(\alpha^2 + \omega^2)} \tag{6.4}$$

It clearly shows that the process has most of its variance at low frequencies, thereby justifying the name red noise process. The "stochastic climate model" [58] predicts the emergence of spectra like (6.4) due to internal dynamics.

Observed time series of sea-surface temperature, atmospheric heat flux and sea-ice were found to be consistent with such a red noise process (see the review by Frankignoul [39]; see also Sect. 6.2.1) and demonstrate the power and generality of this stochastic climate model.

Here we present an example, drawn from [115], that shows the stochastic climate model at work in a quasi-realistic ocean model. A quasi-realistic ocean model exposed to white noise fluctuations indeed varies as predicted by the stochastic climate model. Specifically, a time-variable freshwater flux was superimposed on the mean fresh water flux. This time-variable flux was randomly drawn from a white noise process with zero mean, zero correlation in time but finite correlation in space. The global ocean model responded to this random forcing with pronounced well-organized low-frequency variations in many parameters. A time series of the mass transport through the Drake passage is shown in Fig. 6.6. The second order autoregressive model (6.2) was found to describe many aspects of the variability.

A detailed analysis revealed that an eigenmode of the oceanic circulation had been excited in the model. This mode has a quasi-oscillatory behavior with a period of about 350 years, involves the entire meridional circulation of the North Atlantic Ocean, and extends into the Antarctic Circumpolar Current. Its spectrum is also shown in Fig. 6.6. As expected, it is almost red, apart from the spectral peak at the eigen-frequency of about 350 years.

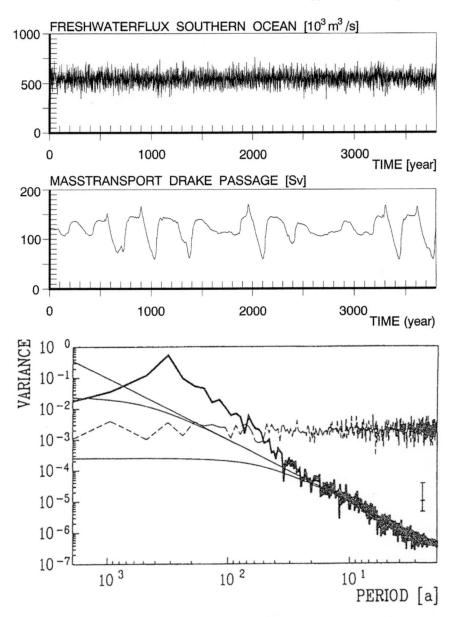

Fig. 6.6. Stochastic climate model at work: *Top:* The white-noise freshwater flux added to a quasi-realistic oceanic general circulation model. *Middle:* Mass transport through the Drake Passage in a numerical experiment forced with the white-noise freshwater flux. *Bottom:* Spectrum of the mass transport through the Drake Passage. For comparison spectra of red-noise processes are also shown. The flat spectrum (*dashed line*) is the spectrum of the white noise forcing. From Mikolajewicz and Maier-Reimer [115]

Later analyses showed that the model results are not realistic. They are due to the specific setup of the experiment and of the model. It has been speculated that this behavior is due to the fact that the model was not coupled to a quasi-realistic atmospheric model but run with prescribed energy fluxes at the ocean's surface. The behavior of the model is thus probably not a positive analog of the real ocean.

6.1.4 Validating Stommel's Theory of the Thermohaline Circulation

Henry Stommel [161] theoretized that the North Atlantic thermohaline or meridional overturning circulation[8] would be sensitive not only to changes in the meridional temperature contrast but also to changes in the freshwater flux pattern. His conceptual model consists of two ocean boxes, one represents the low latitudes and the other one the high latitudes[9]. The overturning circulation is represented as a heat and salt exchange term between the boxes. The basic equations are heat and salt conservation, and are given by

$$\frac{d}{dt}\Delta T = \lambda(\Delta T^* - \Delta T) - 2|m|\Delta T$$

$$\frac{d}{dt}\Delta S = 2\Delta F - 2|m|\Delta S$$

(6.5)

where $\Delta T > 0$ is the positive South–North temperature difference and ΔS the normally negative South–North salinity difference. The first term on the right-hand side of both equations describes the forcing. It is given by "mixed boundary conditions". The heat flux is parameterized as a restoring term, restoring to the temperature ΔT^*. The restoring parameter λ is large, so that the surface flux produces a nearly fixed temperature difference between the two boxes with $\Delta T > 0$. The fresh water forcing is a prescribed flux ΔF, which should be inferred from the atmospheric water vapor flux from low to high latitudes. The second term in both equations describes the transport between the boxes. The transport velocity m is assumed to be proportional to the density difference[10] between the two boxes:

$$m = k(\alpha\Delta T - \beta\Delta S)$$

(6.6)

---

[8] The North Atlantic overturning circulation is part of the conveyor belt (see Fig. 1.12) with a northward inflow of warm surface water across the equator to the northern North Atlantic, a densification and sinking of this water in the northern North Atlantic and southward return flow in the deep Atlantic.

[9] The following version of Stommel's box model was first given and solved by Marotzke [111]. Our discussion is based on [183].

[10] For simplicity, the equation of state is linearized.

For $m > 0$ the surface transport is northward, in the direction of the normal overturning circulation whereas for negative m the surface transport is southward, opposite to the normal overturning circulation.

The equilibrium solutions of the system (6.5) and (6.6) can be derived analytically. Combining the steady state of the second equation of (6.5) with (6.6) results in the equation

$$m|m| - \alpha k \Delta T |m| + k\beta \Delta F = 0 \tag{6.7}$$

Its solutions for a northward surface transport, $m > 0$, are

$$m = \frac{\alpha k \Delta T}{2} \begin{cases} +\sqrt{\frac{(\alpha k \Delta T)^2}{4} - \beta k \Delta F} & \text{for} \quad \Delta F < \Delta F^{crit} \\ -\sqrt{\frac{(\alpha k \Delta T)^2}{4} - \beta k \Delta F} & \text{for} \quad 0 < \Delta F < \Delta F^{crit} \end{cases} \tag{6.8}$$

and its solution for a southward surface transport $m < 0$ is

$$m = \frac{\alpha k \Delta T}{2} - \sqrt{\frac{(\alpha k \Delta T)^2}{4} + \beta k \Delta F} \quad \text{for} \quad 0 < \Delta F \tag{6.9}$$

Equations (6.8) and (6.9) represent different branches of equilibrium solutions and are shown in the left of Fig. 6.7. Two of them, indicated by solid lines, are stable, and one of them, indicated by the *dashed line*, is unstable.

For $\Delta F < 0$, (6.8) describes a thermohaline driven flow, since both the temperature difference ΔT (cooling of northern water) and fresh water forcing ΔF (saltening of the northern box) work together to increase the density of the water in the northern box and to intensify the overturning flow.

For $\Delta F > 0$, the northward flow given by (6.8) is purely thermally driven. Freshening the northern box slows down the overturning flow m. With increasing ΔF, m decreases. This solution branch ends in a bifurcation (point S) at a critical fresh water input of $\Delta F^{crit} = (k\alpha^2 \Delta T^2)/(4\beta)$. The second equilibrium solution with $m > 0$ is also thermally driven (*dashed line* in Fig. 6.7a). It is not only much weaker than the other solution, but it is also unstable. The southward flow solution $m < 0$ (6.9) is purely haline driven ($\Delta F > 0$).

When the model (6.5,6.6) is integrated forward in time with a fixed ΔT^* and slowly varying ΔF, the solution follows the light curves in Fig. 6.7. Start the integration on the thermohaline branch with a negative ΔF. When one slowly increases ΔF this northward flow solution is maintained until the freshwater flux approaches the critical value ΔF^{crit}. When the freshwater flux is further increased, the surface transport drops quickly until it reaches the southward, haline-driven solution. When the freshwater flux is then reduced, the transport does not return to its old values, but remains on the haline branch, until it jumps back to a strong northward surface flow, when the flux becomes negative. The response of the system to changing freshwater fluxes exhibits a *hysteresis*.

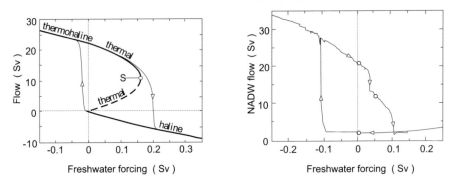

Fig. 6.7. Stommel's hysteresis of the overturning circulation derived from a conceptual model and from a simulation with a quasi-realistic ocean model. *Left:* Equilibrium flow m given a freshwater flux ΔF (*solid line*) and results of forward integration (*light line*). The arrows indicate whether ΔF is increased or decreased during the integration. *Right:* Intensity of the overturning flow in Rahmstorf's forward integration of a quasi-realistic ocean model with gradually changing freshwater flux F. Again, the arrows indicate the direction of the change of F. For more details see [136]

The question is, of course, whether this conceptual model bears any similarity with the real world dynamics. To investigate this question Rahmstorf [136] designed a numerical experiment for the Atlantic with a quasi-realistic ocean model[11]. Similarly to the integration of the conceptual model (6.5,6.6), Rahmstorf integrated the GFDL general circulation ocean model with mixed boundary conditions, a fixed restoring temperature T^* and a slowly changing freshwater input F at high latitudes. First the freshwater flux was slowly increased, from a negative initial value, until reaching the breakdown of the overturning circulation for some positive ΔF-value. Then the freshwater flux was slowly decreased again, until recovering the northward surface flow for some negative flux value. The result is shown in the right panel of Fig. 6.7. The response is again a hysteresis and very similar to that obtained from the box model shown in the left panel of Fig. 6.7.

Rahmstorf's [136] experiment suggests that the three-dimensional overturning circulation possesses *multiple equilibria* and hysteresis similar to the simple box model. It should, however, be realized that Fig. 6.7 compares an intrahemispheric box model with an interhemispheric general circulation model and that the circulations in the two models are not identical. The similarity of the two panels might thus be fortuitous. Interhemispheric box models will eventually provide a more accurate conceptualization of the sensitivity of the

---

[11] Rahmstorf [136] is not the first GCM study of multiple equilibria and hysteresis of the thermohaline circulation. The first study of multiple equilibria with an OGCM was by Bryan [15] and with a coupled AOGCM by Manabe and Stouffer [108]. The first study of the hysteresis was by Stocker and Wright [159]. More on the early history of the modeling of the thermohaline circulation can be found in [112].

Atlantic thermohaline ciculation. Nevertheless, Stommel's visionary hypothesis will remain at the heart of all these conceptualizations[12].

6.1.5 Validating Theories of Alpine Lee Cyclogenesis

Egger [33] and Tafferner and Egger [165] devised a method to test the validity of theories that claim to have identified the first order processes responsible for a certain phenomenon. They apply their method to Alpine lee cyclogenesis, i.e., to the formation of small intense cyclones in the lee of the Alps. The basic idea is to first simulate the cyclone formation in a quasi-realistic model, and then to constrain the model so that it satisfies the assumptions of the theory to be tested. If cyclone formation also takes place in the constrained model, similarly to the unconstrained model, then the theory is accepted, at least for the case considered[13].

Three theories have been suggested for Alpine lee cyclogenesis. They are all linear and assume a basic mean flow. Two of the theories assume the mean flow to be zonal. Cyclogenesis is then ascribed to the interaction of an initial baroclinic wave perturbation with the mountain range. The mean flow is not affected by the orography. Thus, both theories essentially claim that lee cyclogenesis can be understood as a growing baroclinic wave, modified by mountains. The two theories differ in their treatment of the orography. In one theory, the orography is described as an infinitely narrow barrier blocking all meridional flow in the lower levels of the model. In the other theory the mountains are given as a shallow, smooth obstacle. The third theory does not need any initial disturbance but requires a non-zonal mean state with significant vertical shear. Under these assumptions a standing baroclinic wave grows in the lee of the mountain range. For further details and references see [165].

In a first attempt, Egger [33] used an idealized "quasi-realistic model" with a barotropic low pressure system located northwest off the Alps embedded in a westerly mean flow with vertical shear. In the model, the low pressure system moves eastward, and a small but intense cyclone forms in the lee of the Alps within 24 hours. Then, the assumptions of the above three different linear theories were imposed on the model. For example, the Alpine topography was replaced by a razor blade for theory one. In none of these constrained models cyclogenesis occurred similar to that in the "quasi-realistic" model. Thus the theories and their core processes do not seem to have any relevance.

---

[12] Historically, Stommel's 1961 paper which is now referred to as seminal was not seminal at all but ignored for more than twenty years. It was the box model of Rooth [143] and the personal contact between Claes Rooth and Frank Bryan that led to the first GCM experiment on multiple equilibria [15]. Stommel's paper was then rediscovered (Jochem Marotzke, pers. comm. and [112]).

[13] There is no reason to believe that there is only one type of process leading to the formation of lee cyclones.

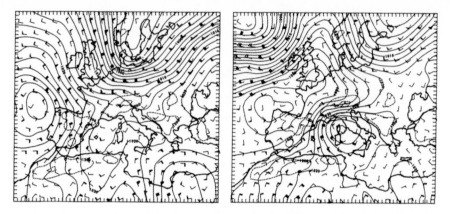

Fig. 6.8. Observed sea level air pressure (spacing 2.5 hPa) and near surface wind distribution at 12UTC March 4, 1982 (*left*) and 12UTC, March 5, 1982 (*right*). From Tafferner and Egger [165]

It was, however, speculated that this failure was due to the idealized "quasi-realistic" model and to the idealized initial conditions. The quasi-realistic model and initial conditions were not realistic enough. Therefore the analysis was repeated, this time with a *quasi-realistic regional* atmospheric model and with two initial synoptic situations that were observed to generate Alpine lee cyclones [165]. Similar results were obtained for both cases. Here we discuss only one case, the 24 h development beginning on 12UTC, March 4, 1982. It was the strongest cyclonic development during the ALPEX observational campaign[14]. The observed surface pressure maps together with surface wind fields for 12UTC on March 4 and March 5 are shown in Fig. 6.8. The episode began with a southwesterly flow at upper levels, and weak pressure differences in the Alpine region. The upper air trough moved eastward, and a marked low pressure center formed over Corsica.

The regional model is initiated with the conditions at 12UTC, March 4 and integrated for 24 hours. The eastward movement of the upper air trough is correctly simulated, and the small scale low is developing almost as observed (see Fig. 6.9a). It is therefore concluded that the model reproduces the observed phenomena in a satisfactory manner.

When the quasi-realistic regional model is constrained to the assumptions of the three theories and integrated for 24h, then the pressure and wind fields shown in Fig. 6.9b-d are obtained. Obviously, in none of the constrained simulations a small scale cyclone like the one in Fig. 6.9a is formed. From this much more stringent test it is concluded that the three theories have indeed not identified the dominant processes for the formation of the lee cyclone. They may be mathematically and conceptually intriguing and appealing, but they do not stand the test of being practically relevant.

---

[14] An international campaign, from September 1, 1981 until September 30, 1982.

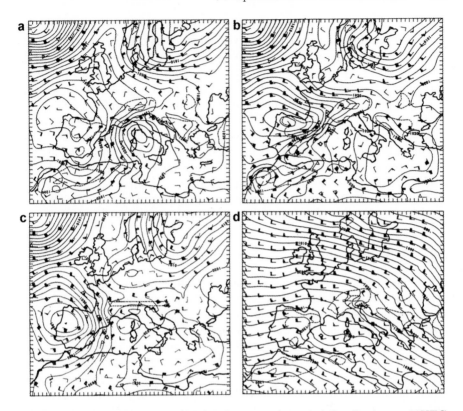

Fig. 6.9. Sea level air pressure (hPa) and near surface wind distribution on 12UTC, March 5, 1982 as simulated with the full model (*upper left*), and with three constrained model versions consistent with three theories to be tested. The razor-blade topography assumed in one of the theories is depicted in the *lower left* as a *dashed box*. From Tafferner and Egger [165]

6.2 Specification of Reduced Models

Here we present two cases where the *output* of quasi-realistic simulation models is used to specify free parameters in *conceptual models*. The functional form of conceptual models is often given, mostly by theoretical considerations, often supported by empirical evidence, but contains free parameters whose values are not determined within the theory. The role of the quasi-realistic model is then first to examine whether the conceptual model describes relevant aspects of reality and, second, to specify the free parameters. This procedure assumes, of course, that the quasi-realistic model represents reality. However, it transforms the output of the model, a huge set of numbers, into *knowledge*.

The first example is drawn from [184] and considers a conceptual model of the interaction of atmospheric heat fluxes and sea surface temperatures. The model is broadly formulated and contains different types of interactions. The output of a very long climate model simulation is used to decide which type dominates in which region of the world ocean and to specify free parameters. In principle, the same procedure could be applied to observational data. However, they are generally too short and any estimation attempt runs into severe uncertainty problems. The very long data sets from a quasi-realistic climate model do not suffer from this problem and yield more definite results.

The second example, drawn from [166], considers a zero-dimensional energy balance model that is capable of describing climate change related to greenhouse gases. The model is calibrated by comparison with detailed scenario simulations, done with a quasi-realistic climate model. The calibrated model is both valuable as a conceptual model, encapsulating our knowledge about the sensitivity of global temperature to greenhouse gas concentrations, and suitable for use in simplified climate–economic modeling.

6.2.1 Heat Flux and Sea Surface Temperature

The atmosphere and ocean are coupled through fluxes of momentum, heat, freshwater and gases. The ocean influences the atmosphere above; the atmosphere influences the ocean below. Examples of such air–sea interactions are manifold. Evaporation over warm sea surface temperatures increases the water vapor content of the air above. When such moist air is brought in from the warm Atlantic Ocean to the eastern coast of the United States it is often dumped as a thick layer of snow during blizzards. Conversely, the atmosphere drives the ocean circulation. Both the Gulf Stream in the Atlantic and the Kurushio in the Pacific are wind driven. The seasonal freezing of the Baltic Sea is caused by cold air temperatures and the low salinity of the Baltic Sea. These mutual influences, together with other atmospheric and oceanic processes, establish the mean state of the atmosphere–ocean system. The question is what happens when this mean state is perturbed by *anomalies* in either system. The answers depend on the time scale considered.

In the following we consider times scale of months and seasons. On these time scales the major coupling between the ocean and atmosphere is through the heat flux H which is affected by the sea surface temperature T which in turn is affected by the heat flux. Again, the question is how anomalies feed back on each other. To entangle the various possible feedback processes J. von Storch [184] followed the approach of Frankignoul [40], [39] and decomposed the heat flux into the ocean H into

$$H = H_o + H' + H^* \qquad (6.10)$$

where H_o represents the long term (seasonal) mean, H' the anomalies due to internal processes in the atmosphere and H^* the anomalies caused by air–sea interaction. It is assumed that H', H^* and T vary according to the conceptual model

$$H'_t = \lambda_H H'_{t-1} + n_t^H$$
$$H^*_t = \alpha_H T_{t-t_H} \tag{6.11}$$
$$T_t = \lambda_T T_{t-1} + \alpha_T H'_{t-1} + n_t^T$$

The terms n_t^H and n_t^T are *noise* terms, unrelated to the state of both H and T. Thus, the internal heat flux variations H' are assumed to depend on the previous heat flux, a memory effect, and to be driven by random atmospheric noise. Formally, this is an AR(1) autoregressive process of 1st order. The interactive heat flux variations H^* are assumed to be entirely given by the state of the sea surface temperature t_H time steps earlier. Similarly, the SST variations are assumed to be due to a memory effect in the ocean, the variations in the heat flux H', and random oceanic noise. The values of the parameters $\mathrm{VAR}(n_t^H)$, $\mathrm{VAR}(n_t^T)$, λ_H, α_H, λ_T and α_T determine the strength of the various processes.

As a diagnostic tool J. von Storch considers the cross-covariance function

$$\gamma_{H,T}(\tau) = E[(H_t(t+\tau) - H_o) \cdot (T_t(t) - T_o)] \tag{6.12}$$

The heat flux leads the sea surface temperature for positive τ values. This cross-covariance function can be written as the sum of two terms

$$\gamma_{H,T} = \gamma_{H',T} + \gamma_{H^*,T} \tag{6.13}$$

and be calculated from the conceptual model (6.11). The first term becomes

$$\gamma_{H',T}(\tau) = \begin{cases} \sigma^2 \alpha_T \lambda_T^{\tau-1} \sum_{i=0}^{\tau-1} (\lambda_H/\lambda_T)^i & \text{for } \tau > 0 \\ 0 & \text{for } \tau \leq 0 \end{cases} \tag{6.14}$$

where σ^2 is the variance of H'. It vanishes for negative τ when SST leads the heat flux. This cross-covariance function is plotted for two sets of parameters in Fig. 6.10 (*dashed curves*).

The second term becomes

$$\gamma_{H^*,T}(\tau) = \alpha_H \gamma_T(\tau + t_H) \tag{6.15}$$

where γ_T is the (symmetric) auto-covariance function of T. The cross-covariance function $\gamma_{H^*,T}(\tau)$ is thus a symmetric function relative to $-t_H$. The *solid curve* in Fig. 6.10 is an example of this type of cross-covariance function, with $t_H = 2$ months.

The total cross-covariance function is thus a sum of a function that is symmetric to some time lag, and a function that is zero for negative lags. Functions like the dashed-dotted ones in Fig. 6.10 result from this combination. A cross covariance function like the *thin dashed-dotted line* suggests that a, say, positive anomaly will persist for an extended period of time. The fact that the signal is transferred from the atmosphere to the ocean and back,

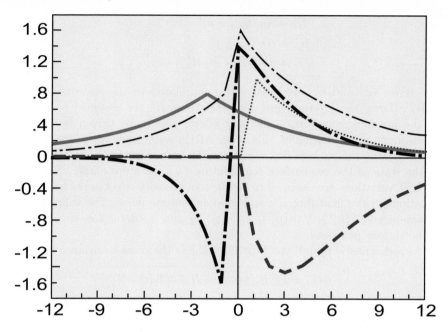

Fig. 6.10. Theoretical cross-covariance functions for different interactions of heat flux and sea surface temperature. The *dashed curves* describe situations where heat fluxes cause SST variations, as in (6.14). The *solid curve* is characteristic for the reverse situation, where SSTs cause heat flux variations, as in (6.15). The two *dash-dotted curves* are sums of curves of the solid and dashed type. The *thin dash-dotted curve* represents a weak positive feedback and the *thick dash-dotted curve* a strong negative feedback. Adapted from J. von Storch [184]

does not change the signal significantly. This interaction is called a *weak positive feedback*. The *thick dash-dotted curve* in Fig. 6.10 describes a different situation. It changes its sign at zero lag. An initially positive anomaly is transformed into a negative anomaly after some time, through the exchange with the other system. This interaction is called a *strong negative feedback*.

To test the validity of the above conceptual model and to determine its parameters, cross-covariance functions were derived from a 300-year simulation with a regular climate model. The climate model consisted of a coupled atmosphere–ocean general circulation model. The main advantage of using such a model is that the estimation of the cross-covariance function becomes rather robust. This is demonstrated in Fig. 6.11. When only 20 years of monthly data are used, the cross-covariance function varies irregularly. Only when much more data, in this case 300 years, are used, robust features clearly emerge. This is the major advantage of the climate model. Observational data cover at best a few decades. This is not long enough to reliably estimate and conceptualize the air–sea interaction.

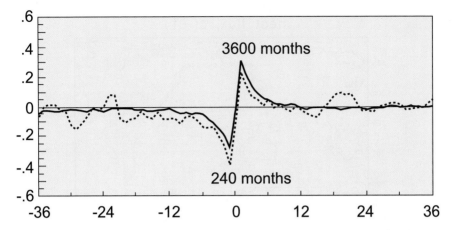

Fig. 6.11. Estimated cross-covariance functions of simulated heat flux and sea surface temperature. The relatively smooth *solid curve* is determined from a sample of 3600 months (300 years). The more irregular *dotted curve* is based on only 240 months (20 years) from the same model simulation. From J. von Storch [184]

The cross-covariance function shown in Fig. 6.11 is typical for extratropical conditions in the Pacific. It describes air–sea interaction with a strong negative feedback. When, due to the presence of, say, anomalously warm air, a downward (positive) heat flux exists at some time, then the sea surface temperature is increased, with the effect that the air–sea temperature difference is reduced and even an upward anomalous heat flux may emerge, when the air temperature returns to normal conditions.

The cross-covariance function $\gamma_{H,T}(\tau)$ has been estimated for the lags $\tau = -1$, zero and $+1$, for all ocean grid points. The values are plotted in Fig. 6.12. Over most of the global ocean, with the exception of the tropical ocean and polar regions, the cross-covariance function is very small for $\tau = -1$, i.e., when the SST leads. Larger values emerge for zero lags and for lags $\tau = +1$, when the heat flux leads.

Two characteristic numbers are determined at each ocean grid point. The first number indicates the relative importance of H and T. It is the lag at which the cross-covariance function has its maximum. When this lag is positive, then likely the heat flux H leads the sea surface temperature T. A negative lag points to the sea surface temperature leading H. The result is shown in Fig. 6.13. According to the climate model and the interpretation by the conceptual model, SST variations are almost everywhere a result of heat flux variations. The opposite link is only found in the tropical ocean and in parts of the polar oceans, where sea ice formation and dynamics complicate the situation.

The second characteristic number is the ratio $r = \gamma_{H,T}(1)/\gamma_{H,T}(-1)$. When a heat flux generates SST variations and the heat flux itself is not

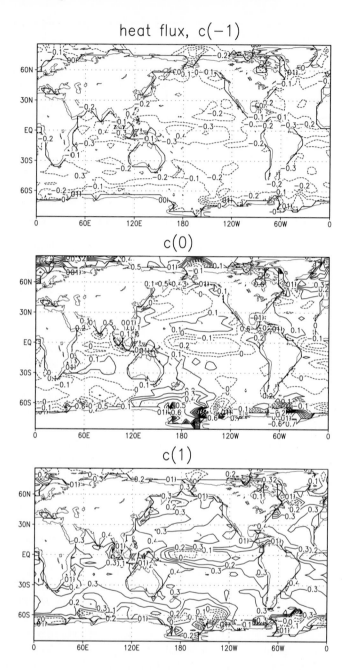

Fig. 6.12. Cross correlation functions between heat flux and sea surface temperature at lags −1, 0 and +1, derived from a 300 year simulation with a quasi-realistic climate model. From J. von Storch [184]

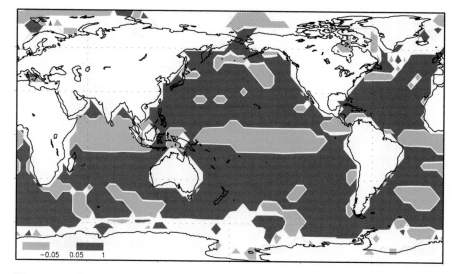

Fig. 6.13. Time lag at which the cross covariance function of heat flux and sea surface temperature is maximum. Positive (*dark*) lags indicate a lead of the heat flux, and negative (*light*) lags a lead of SST. From J. von Storch [184]

changed, then the cross-covariance function is solely given by (6.14). The denominator of r becomes zero so that $r = \infty$. When a SST anomaly causes a heat flux anomaly that does not change the SST then the numerator of r will be zero so that $r = 0$. A weak positive feedback is associated with a positive $r = 0(1)$ because both the numerator and denominator are comparable and have the same sign. A strong negative feedback is associated with negative r values. The example shown in Fig. 6.11 displays a case of negative r.

The number r is displayed for all ocean grid points in Fig. 6.14. Almost everywhere, the ocean is covered with moderate negative r-values. Such values indicate a regime where an original SST or heat flux anomaly is eventually damped by strong negative feedback. In the Central Tropical Pacific near-zero values of r prevail, indicating a regime where SST strongly influences the heat fluxes without any feedback. In the polar regions different situations are found.

The conceptual model (6.11) and its calibration by comparison with a coupled atmosphere–ocean general circulation model does not identify the *physical processes* responsible for air–sea interaction but it does identify and classify different air–sea interaction regimes. This is a useful result for applications. Instead of running an expensive coupled atmosphere–ocean general circulation model when investigating other questions, one can run a less expensive ocean general circulation model forced with heat fluxes determined by the model (6.11). The atmospheric dynamics is encapsulated in the parameters of the model.

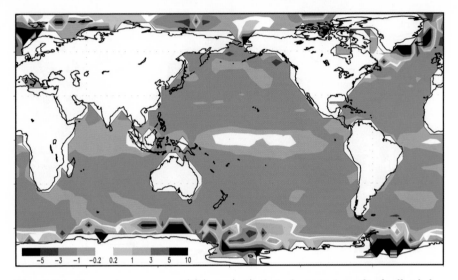

Fig. 6.14. Parameter $r = \gamma_{H,T}(1)/\gamma_{H,T}(-1)$ that characterizes the feedback between SST and heat flux. For color scale see inset. From J. von Storch [184]

6.2.2 A Conceptual Zero-dimensional Climate Model

The following example, taken from [166], demonstrates how a complex data set, generated by a quasi-realistic climate model, may be used to derive a conceptual model – in this case an energy balance model that describes variations of the global mean temperature. This "zero-dimensional" model is again useful since it encapsulates our complex knowledge about the sensitivity of climate to greenhouse gas forcing. It has also been used as a convenient tool in the design of climate change scenarios and adaptation and mitigation policies [166], [125].

The concept of the energy balance and a simple energy balance model were already introduced in Chaps. 1 and 4[15]. The state variable of such energy balance models is the global mean air temperature near the surface of the earth. A particular model is

$$\frac{d}{dt}T(t) = -\lambda T + \mu C + n \qquad (6.16)$$

where C represents the deviation of the global mean concentration of greenhouse gases from the pre-industrial level, and T the deviation of the global mean air temperature from an equilibrium temperature consistent with the pre-industrial concentrations. The λ-term is a restoring term. It describes the tendency of the system to return to its equilibrium temperature, when the concentrations return to "normal" (i.e., to the pre-industrial value). The

---

[15] See in particular Figs. 1.11 and 4.9.

"source"-term μC describes the heating effect of increased greenhouse gas concentrations[16]. Variations generated within the climate system are accounted for by the noise term n.

An open question is whether (6.16) is a reasonable description of the real world, and, if so, what the values of the unknown parameters λ, μ and $\text{VAR}(n)$ are. Ideally, one would test (6.16) and specify the parameters with observed data. This is possible only to a limited extent. Temperature data have been recorded only since about 1860, and during this time the variations of greenhouse gas concentrations have been moderate. A viable alternative is to consider the output of a quasi-realistic climate model run with increasing greenhouse gas concentrations as described in Sect. 5.3.

Tahvonen et al. [166] used (6.16) and fitted the constants λ and μ to the output of a "scenario"-simulation by Cubasch et al. [21]. The model simulated the response to a prescribed increase of greenhouse gas concentrations. The greenhouse gas concentrations were specified in terms of "equivalent CO_2 concentrations" according to the Scenario IS92a[17] of the IPCC, with 335 ppm in 1985 and 627 ppm in 2085. The fit of (6.16) to the simulated data, using annual means and some smoothing, gave $\lambda = 2.6 \times 10^{-2}/\text{year}$, and $\mu = 3.9 \cdot 10^{-4}\,\text{ppm}/(\text{yearK})$.

A comparison of the model predictions with observations is shown in Fig. 6.15. The thin curve shows the model temperature when it is calculated from (6.16) using the two numbers above, no noise term and the time variable concentrations C between 1860 and 1987. It reproduces the observed trend reasonably well. Part of the observed year-to-year variability is "random" and could be modeled by the noise term. Another part is due to systematic variations in solar output and in tropospheric aerosols. In principle, these systematic effects can also be included by adding suitable terms to the conceptual model (6.16).

For comparison, the constants λ and μ were also fitted to the observed data from 1860 to 1987. Interestingly, very similar numbers are obtained, and when inserted into (6.16), a very similar trend for the temperature results (*thick line* in Fig. 6.15). This result gives credibility both to the original scenario simulation and to the conceptual model (6.16).

The conceptual model (6.16) allows, in a straightforward manner, the estimation of the climatic response to various scenarios of the emissions of greenhouse gases into the atmosphere. In conjunction with economic models the model can thus be used, and has been used, to evaluate different policies that aim at a combination of abatement of and adaptation to climate change. One can perform cost-benefit analyses that identify "optimal" policies, which min-

---

[16] A better approximation is to use $log(C)$ instead of the linear term C, but for small concentrations this difference is immaterial. In the study by Tahvonen et al. [166] the linear term was used to facilitate a subsequent optimization problem involving economic costs and benefits.

[17] See Fig. 5.36.

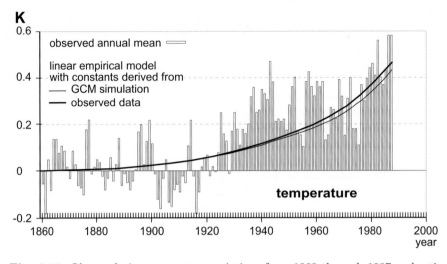

Fig. 6.15. Observed air temperature variations from 1860 through 1987 and estimated temperature increase, calculated from (6.16) as a response to the observed increase of greenhouse gas concentrations from 1860 to 1987. From Tahvonen et al. [166]

imize the abatement costs given some climatic thresholds, or minimize the total abatement and adaptation costs [166].

6.3 Simulating the Unobservable

Any analysis of the environment is hampered by the fact that the observational record is both incomplete and limited. Only a few variables are recorded at a limited number of positions. Subtle but important aspects, like the deviation from uni-modal distributions, can often not be inferred from such observational data. Such incomplete observations may be interpolated in a dynamically consistent way by assimilating them into a quasi-realistic model (as is, for instance, done in the NCEP/NCAR global re-analysis of atmospheric data – see Sect. 5.2.1) but this interpolation usually does not recover relatively subtle aspects either. The only alternative in these cases is to trust the output of quasi-realistic models. We present three examples:

- statistics of circulation regimes in the North Sea,
- multiple equilibria of the atmospheric circulation, and
- tidal dissipation.

Other examples are the analysis of the global energy cycle [103], [168], and the analysis of empirical links between proxy-data and climate variable [195]. The latter analysis aims at determining those links that remain stationary over hundreds of years and can be used for the inversion of proxy data.

6.3.1 Circulation Regimes in the North Sea

A straightforward example of constructing knowledge about something unobservable is the case of the statistics of the flushing circulation of the North Sea in Northwest Europe. From a variety of limited observations, a number of different circulation regimes were identified in the North Sea [4], varying on time scales of days and weeks. The most frequent regime is one with a counterclockwise rotating circulation cell, but sometimes the circulation is clockwise, less organized and even ceases altogether.

Unfortunately, statistics about the frequency of the different regimes and the transition probabilities between the different regimes, including the persistence of remaining in a regime, cannot be derived from observations since no long-term daily analyses of the North Sea circulation exist.

Therefore, Kauker [78] integrated a quasi-realistic North Sea model for 15 years, forced with the atmospheric fluxes of the ECMWF re-analyses[18]. The model output was compared with observations of water level at coastal stations and temperature and salinity maps, which are available for some months. The comparison was satisfying, and the model considered validated. The observations were found to be positive analogs . The neutral analog "daily circulation" was analyzed in some detail [79]. Only winter conditions were considered, one field per day in the months December, January and February.

Five different circulation regimes were found[19]. Typical streamfunction patterns for these regimes are shown in Fig. 6.16. Regimes I and III exhibit typically two opposite one-cell circulations, regimes II and IV are somewhat reminiscent of opposite double cell patterns, and regime V exhibits almost no circulation. The most frequent regime is the counterclockwise rotating one cell regime III with a frequency of 30%; the second most frequent regime is regime II with a frequency of 29%. Regimes I and IV appear with a probability of about 15%, and 10% of all cases are classified as belonging to the "ceased" circulation regime V. The frequencies of transitions between the different regimes "X" and "Y", abbreviated by X →Y, are shown in Table 6.1.

Thus, the system has a clear tendency of remaining in the present regime (persistence), but also a weak tendency towards a "circular" evolution " $\cdots$ →I →II →III →IV". This feature is related to the climatological characteristic of (cyclonic) storms passing across the North Sea in easterly direction.

---

[18] A similar simulation was done by Janssen [70] for the Baltic Sea.

[19] Formally, these regimes represent five different subsets of a two-dimensional phase space. The phase space is spanned by the first two EOFs of daily streamfunction variations. For details, refer to [79].

Table 6.1. Frequencies of transitions between regimes "X" and "Y"

	Y					
X	X→I	X→II	X→III	X→IV	X→V	sum
I→Y	10.6	2.6	0.1	0.4	0.6	14.4
II→Y	1.2	20.1	4.8	0.6	2.2	28.9
III→Y	0.1	2.6	23.1	3.8	0.6	30.2
IV→Y	1.5	0.8	1.7	8.4	2.5	14.8
V →Y	1.0	2.7	0.6	1.6	3.6	9.5

This type of analysis could only be done with the help of a model. In this case, the output of a model was used that was validated by the available observations. Certainly, the output of a model that assimilates the available observations would be superior and ensure a better similarity of the modeled and real ocean state, but such multi-year oceanic re-analyses have not been done yet.

6.3.2 Multimodality in Atmospheric Dynamics

The atmospheric dynamics on time scales of days, weeks and longer are nonlinear, a fact clearly documented by the limited success of predicting weather for more than a week or so. In order to understand this nonlinear behavior, Edward Lorenz [102] reduced the atmospheric dynamics to a system of three ordinary differential equations. This system exhibits chaotic behavior, the famous butterfly effect. It has two distinct attractors and flips irregularly between the two. Many efforts were launched to identify such attractors or stable states in the large-scale atmospheric circulation. One prominent example is the theory of Charney and DeVore [18]. They derived a conceptual model of the extratropical circulation that exhibits two distinctly different circulations. One circulation is almost zonal. The stream function is east-west banded. The other circulation is "blocked". The stream function is strongly wavy. These two states are reminiscent of the two major winter weather situations in the extratropics, namely a continuous westerly flow with its associated sequence of extratropical storms, and a blocking situation, where a high pressure system prevails over, say, Europe and "blocks" storms from entering Europe. Thus, it was thought that Charney and DeVore's mechanism would help to understand and eventually predict this phenomenon. And meteorologists started to search for the fingerprint of this mechanism in observational data. They searched for the presence of multimodal distributions of large-scale atmospheric variables.

An early success was claimed by Hansen and Sutera [52], but a subsequent analysis by Nitsche et al. [124] demonstrated that the present amount of observational data is not sufficient to detect multi-modality in two or more dimensions and that much more data are required than are presently available.

This detection problem has been addressed systematically by Berner and Branstator (pers. comm.) using a quasi-realistic general circulation model of

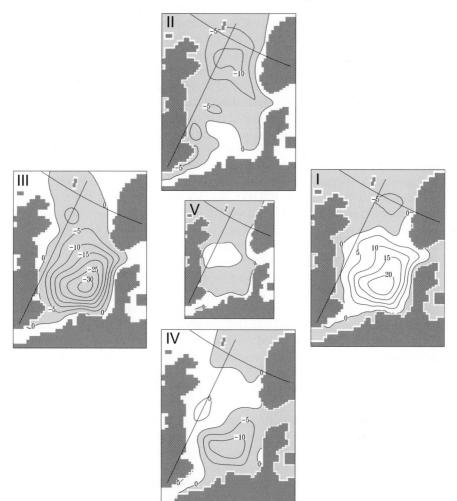

Fig. 6.16. Typical surface velocity streamfunction in the five regimes of the North Sea Circulation. Negative values are shaded. Units: $10^3 m^2/s$. From Kauker and von Storch [79]

the atmosphere. They ran the model repeatedly, from slightly different initial conditions, for a very long time under "perpetual" January conditions[20]. They analyzed 20 simulations of 50,000 days each. With data stored twice daily, each

---

[20] "Perpetual" January simulations were popular in the 1980s. They allow the generation of very long stationary data sets. An atmospheric GCM is run under perpetual January conditions by fixing the sea surface temperature distribution, the soil temperature and moisture and the solar insolation to observed conditions in January.

of the 20 simulations generated 100,000 atmospheric states. Thus, a total of 2 million daily atmospheric states were generated to test various hypotheses, among them, the hypothesis of multiple equilibria.

Here we consider the joint 2-dimensional distribution of two EOF coefficients[21]: EOF1 and EOF4 of the 500 hPa global geopotential height. EOF1 is sometimes related to the Arctic Oscillation. EOF4 was chosen because it was found to have the most non-Gaussian distributed coefficients among the leading EOFs. The distributions are estimated by either simple binning (30 × 30 bins) or by using a more sophisticated kernel method, which fits a sum of smooth function to the distribution.

Figure 6.17 displays the estimated distributions derived for different samples of monthly data. When 100,000 monthly maps are used (*top*), then the binning and kernel distributions are rather similar. They are both slightly skewed, with only one maximum. When a smaller number of samples is used, then some of the estimated distributions become bi-modal whereas others remain uni-modal (i.e., have one maximum).

Examples for two different samples of 125 monthly maps are shown in Fig. 6.17. The estimates using the binning method (*middle row*) differ markedly from that using the kernel method (*lower row*). The binning estimate shows in both cases a bi-modal distribution, which has little similarity with the best estimate based on 100,000 months. However the irregular appearance of the distributions warns the expert that there may be an estimation problem. The situation is different for the distributions estimated with the kernel method. Both estimates are smooth, but one of them is also a distinctly bi-modal.

The sample size of 125 monthly maps was, of course, not chosen arbitrarily. It corresponds to the sample size presently available from observations, 3 months per winter for 40 years gives 120. The lesson to be learned from this study is thus that the limited sample size of our observational record may lead to severe misjudgements. Berner and Branstator's very large sample of 100,000 monthly maps, constructed with a *model*, enabled them to determine the "true" distribution with reasonable accuracy, and to demonstrate what type of erroneous assessments become possible when the limited observational evidence available to us today is used.

6.3.3 Tidal Dissipation

How and where the tides dissipate their energy is an open question with consequences for many oceanographic problems. Traditionally, it has been assumed that most of the tidal dissipation occurs through bottom friction in shallow seas but recently it has been suggested that a substantial fraction

---

[21] EOFs are statistically determined characteristic patterns, which are most efficient in describing space-time variability. The coefficients are normalized to a standard deviation of one. For further details refer to Appendix C.2.2.

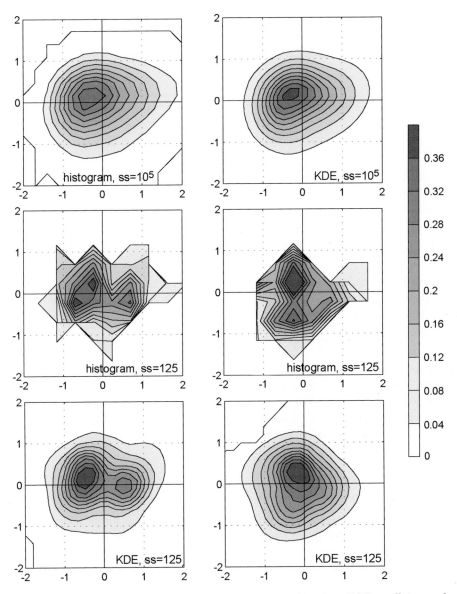

Fig. 6.17. Estimated two-dimensional distributions of leading EOF coefficients of geopotential height derived from simulated data sets. *Top:* Estimation based on 100,000 monthly maps (estimated through binning and with a kernel method). *Middle:* Estimation based on two subsets of only 125 monthly maps each, using binning. *Bottom:* Estimation based on the same two subsets of 125 monthly maps shown above, but using the kernel method. Courtesy of Berner and Branstator

may also dissipate in the deep ocean through conversion of barotropic into baroclinic tidal energy at bottom topography.

The kinetic and potential tidal energies per unit surface area are defined by

$$E_{kin} = \frac{1}{2H_0}\rho_0 \mathbf{U}_h \cdot \mathbf{U}_h$$

$$E_{pot} = \int_{-H_0+\xi_e+\xi_l}^{\xi_g} dz\rho_0 gz = \frac{1}{2}\rho_0 g[\xi_g^2 - (-H_0 + \xi_e + \xi_l)^2]$$

where ξ_e and ξ_l are the bottom level variations due to the earth tides and the tidal load. The energy equation can be obtained by multiplying the momentum equation (4.12) by $\rho_0 \mathbf{U}_h/H_0$ and the volume balance equation (4.8) by $\rho_0 g\xi$ and adding the two. The result is

$$\partial_t(E_{kin} + E_{pot}) + \nabla_h \cdot \mathbf{P}_h = W - F - D$$

where $\mathbf{P}_h$ is the horizontal energy flux vector, W the work done by the gravitational forces, F the energy flux into the bottom, and D the dissipation.

When averaged over a tidal cycle the tendency term vanishes and one obtains

$$\nabla_h \cdot <\mathbf{P}_h> = <W> - <F> - <D> \tag{6.17}$$

where cornered brackets $<\cdot>$ denote the average over the tidal cycle.

When considering global averages, the dissipation can be estimated as the difference of the work done by the gravitational forces W and the energy flux into the bottom F. The horizontal energy flux $\mathbf{P}_h$ only redistributes energy within the ocean. It vanishes when averaged over the ocean. The ocean-integrated tidal energy budget thus reduces to

$$\int_{ocean} <D> = \int_{ocean} <W> - \int_{ocean} <F> \tag{6.18}$$

That part of the total energy input by the gravitational forces which is not balanced by transfer to or from the solid earth, must be taken care of by dissipation in the ocean. The total basin-integrated energy fluxes $\int <W>$ and $\int <F>$ can be estimated from the tidal potential and tidal elevation, resulting in total dissipation rates of about 3.75×10^{12} W or 3.75 TW for all tidal constituents and in about 2.5 TW for the M_2 tide[22].

The question is, of course, *where* this dissipation is taking place. The conventional wisdom is that most of the oceanic tidal dissipation occurs in regions shallower than 200 m. The reason offered for this conclusion is as follows:

---

[22] These oceanic estimates are consistent with estimates for the loss of energy from the earth–moon system which reflects itself in changes of the length of day The energy loss from the earth–moon system is dissipated mostly in the ocean. Only about 5% is dissipated in the solid earth and less than 1% in the atmosphere.

the ratio of the areas shallower than $200\,\mathrm{m}$ to those deeper than $200\,\mathrm{m}$ is about 0.07; tidal currents in shallow water are nearly one order of magnitude larger than those in deep water; since dissipation by bottom friction is cubic in the tidal current speed (for quadratic bottom friction) only a fraction of approximately $0.013 = 0.93 \times 1/0.07 \times 10^3$ is dissipated in the deep ocean. The remainder is dissipated in shallow seas.

This conventional wisdom is supported by contemporary hydrodynamical models. It also seems to hold when altimeter data are assimilated into such models. An example is Kantha et al. [77], who assimilated Topex/Poseidon altimeter data into a high-resolution hydrodynamical tidal model. They evaluated the expression

$$<D> = \rho_0 <\mathbf{F}_h \cdot \mathbf{U}_h> /H \qquad (6.19)$$

with a bottom friction $\mathbf{F}_h \sim \mathbf{U}_h |\mathbf{U}_h|$ (cf., (4.5)) and found that about 0.95% of the M_2 tidal dissipation occurred in shallow seas, consistent with the conventional wisdom[23].

This result should not surprise since the analysis relies on (6.19) which puts dissipation at locations where currents are strongest, i.e., in the shallow seas. The problem is the dependence of the dissipation D on the parameterized friction term $\mathbf{F}_h$. If one computes dissipation from a parameterization then one obtains results consistent with this parameterization. The conventional wisdom is encoded into the model. The data assimilation only improves the estimation of the currents, but not necessarily the estimation of the dissipation.

An alternative approach is to estimate dissipation from the balance (6.17):

$$<D> = -\nabla_h \cdot <\mathbf{P}_h> + <W> - <F> \qquad (6.20)$$

By doing so, one avoids ascribing dissipation to bottom friction. Instead, it is given as the residual.

This alternative approach requires the expressions for the horizontal energy flux

$$<\mathbf{P}_h> = \rho_0 g H_0 <\mathbf{U}_h(\xi + \xi_e + \xi_l)> \qquad (6.21)$$

which is defined by pressure times velocity plus energy density times velocity, the work done by the gravitational forces

$$<W> = \rho_0 <\mathbf{U}_h \cdot \nabla_h \Phi_T> \qquad (6.22)$$

with the gravitational potential (4.11), and the vertical energy flux into the bottom F

$$<F> = \rho_0 g <\xi \partial_t (\xi_e + \xi_l)> \qquad (6.23)$$

which is defined as bottom pressure times the vertical velocity of the ocean bottom. The spatial distribution of the energy flux $< F >$ can be inferred

---

[23] The total dissipation rate was only 2.1 TW, falling short of the required 2.5 TW.

from the tidal potential and elevation whereas the spatial distributions of $<W>$ and $<\mathbf{P}_h>$ depend on the volume transport $\mathbf{U}_h$ and hence require a hydrodynamical model.

Egbert and Ray [32], [31] used this alternative approach and estimated the distribution of tidal dissipation in the ocean for the M_2 tide from equation (6.20). They assimilated Topex/Poseidon data into a hydrodynamical model, allowing for errors in the momentum equation while conserving mass exactly[24]. In this way, they assigned the dissipation to locations where it is "needed" to optimally blend data and model. They used the hydrodynamical model discussed in Sect. 5.2.4 with state variables $\xi(\mathbf{x})$ and $\mathbf{U}_h(\mathbf{x})$. The observation vectors are the differences $d(\mathbf{x}, t)$ of the tidal elevations at the cross-over points of ascending and descending Topex/Poseidon ground tracks with an observation equation similar to (5.8).

Figure 6.18 shows the deduced dissipation rates for two such assimilations, differing in assimilation technique and bottom friction coefficient. The color scale is chosen to emphasize the structures in the open ocean and saturates in some shallow seas where the dissipation generally is highest. There are some blue areas which indicate negative dissipation. This is clearly unphysical but gives a sense of the noise in the maps. Though the maps obtained with the different methods differ in detail they have many features in common. There are basically two areas of high dissipation:

1. The shallow seas (such as the continental shelves, the Hudson Bay, the Yellow Sea, etc.). Here the dissipation is due to the frictional bottom drag.
2. The open ocean ridges (such as the Hawaiian ridge, the Mid-Atlantic ridge, the Tuamotu archipelago, etc.). These are areas where elongated topographic features are perpendicular to the tidal currents.

These topographically relevant areas are encircled in Fig. 6.18 by *solid lines.* To quantify the dissipation patterns Egbert and Ray integrated the dissipation rates over these encircled areas. The integration of (6.20) involves area integrals of the work term $<W>$ and flux term $<F>$ and line integrals of $\mathbf{P}_h$ along the boundaries. By putting the boundaries in deep water they minimized the effect of uncertainties in the volume transports $\mathbf{U}_h$. The dissipation estimates for the shallow seas are thus fairly accurate although tidal currents in shallow seas are generally not. The result of these integrations is shown in Fig. 6.19, including additional cases.

All calculations give a value of 2.44 ± 0.01 TW for the total dissipation rate. Out of this 0.70 ± 0.13 TW occur in the deep ocean. The values for the

---

[24] Estimating dissipation from the energy balance (6.20) requires that one enforces mass balance quite strongly. The reason is that transports are estimated from the momentum balance which contains the gradient of the elevation field. Dissipation estimates then requires another gradient to calculate flux divergences. If mass is strictly conserved then the term involving two derivatives is replaced by $i\omega\xi$ and dissipation really only requires one derivative, making it much less prone to errors.

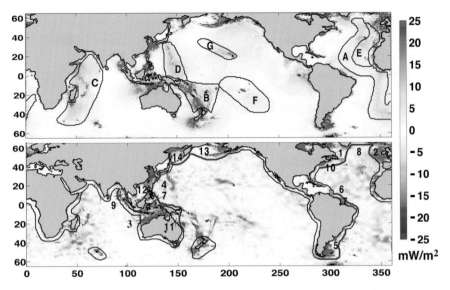

Fig. 6.18. M$_2$ tidal dissipation rates, estimated by combining Topex/Poseidon altimeter data with two different hydrodynamical models in different ways. The *solid lines* encircle high dissipation areas in the deep ocean (*top*) and demarcate the boundary between deep and shallow seas (*bottom*). From Egbert and Ray [32]

hydrodynamical model without data assimilation are 2.02 TW and 0.06 TW, respectively. Assimilation of altimeter data thus causes a considerable change in the location of tidal dissipation: About 25–30% occurs in the deep ocean, as opposed to only 3% in the conventional hydrodynamical model.

The spatial distribution of the deep-ocean dissipation suggests conversion of tidal energy into baroclinic tidal energy at topography. When extrapolated to all tidal harmonics about 1 TW might be dissipated in the deep ocean. This would resolve a long standing oceanographic problem. The abyssal density field of the meridional overturning circulation is maintained by a balance of upwelling and vertical mixing. It is estimated [120] that this mixing requires a mechanical energy input of about 2 TW. About 1 TW is provided by the wind [192]. The tides may provide the other 1 TW. As pleasing as this scenario might be, it must also be realized that the physical processes involved in such a scenario are not fully identified yet, but are under intense study.

Of course, now that the data assimilation exercise has suggested that tides dissipate part of their energy in the deep ocean, presumably by conversion to baroclinic tides at topography, one can introduce this process into the hydrodynamical model – and get a reasonable representation of this process without resorting to data and their assimilation into models. Indeed, a hydrodynamical model that accounts for deep-ocean dissipation by an appropriate drag law produces a dissipation distribution not much different from the data as-

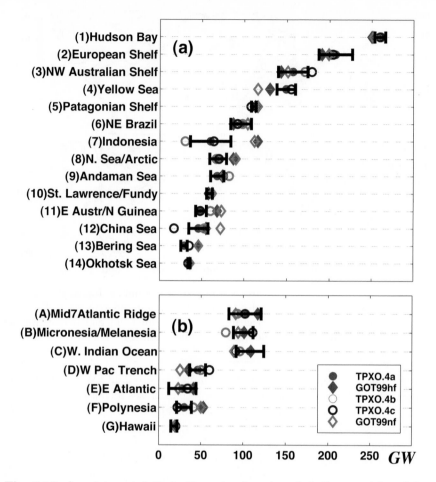

Fig. 6.19. Area-integrated dissipation rates for selected shallow sea (**a**) and deep ocean (**b**) areas. Results are given for five different formulations of the hydrodynamical model and assimilation scheme, together with the error bars obtained by a Monte Carlo approach. From Egbert and Ray [32]

simulation estimate. Dynamical models are continuously updated to include the "latest physics".

It should also be kept in mind that the successful blending of data and models and a reasonable outcome, does not preclude the possibility that other approaches, which combine other data with models in different ways, may lead to equally reasonable but different outcomes. Thus, the above exercise suggests that deep-ocean tidal dissipation is a major contributor to the tidal energy balance, but it has not proven that it *is* in reality. It is still a neutral analog. It needs to be confirmed by actual physical measurements.

7

Issues and Conclusions

In the preceding chapters, we have exposed the general philosophy behind the use of quasi-realistic models in atmospheric and oceanic sciences and illustrated it by many examples. Quasi-realistic models are not only an important but an indispensable part in the tool box of the environmental scientist. They are indispensable as they often constitute the only way to extrapolate beyond the immediate range of experience and to determine characteristics of the state of the system.

As outlined by Navarra [122], there are three main parts in the tool box: reduced concepts or theories, empirical evidence and statistical methodology, and quasi-realistic models and numerical experimentation. Generally, none of these alone is sufficient to generate new knowledge. Theory without information reflecting reality, either generated by observations or quasi-realistic models, is as useless as a statistical analysis of environmental data without quasi-realistic modeling or theoretical considerations guiding the analysis through the maze of infinitely many degrees of freedom. The same is true for the subject of this book, the use of quasi-realistic models: alone, they just generate numbers which are usually as incomprehensible as raw and unprocessed observations. The scientist employing quasi-realistic models needs observations or theoretical concepts to validate the model and to extract useful knowledge from it.

There are other important issues as well. In the introduction to this book, in Sect. 1.7, we listed as key issues:

- reduction of information,
- new and old models,
- trustworthiness,
- model builders and users, and
- social and psychological conditioning.

It is worth now revisiting these issues.

7.1 Reduction of Information

Similar to observational programs, models generate large or even huge amounts of numbers. These numbers contain detailed information about the state of the system. Different from observations, these numbers are obtained under controlled conditions. But the numbers by themselves are of little utility.

This is not to say that huge simulation exercises, like the NCEP's re-analyses of the global weather over the past 50 years (Sect. 5.2.1), are not useful. They are tremendously useful. They are, in fact, used in a large variety of studies as key ingredient (see e.g., Sect. 5.2.2). But the re-analysis itself does not provide new knowledge about atmospheric dynamics. Its value lies in providing an excellent data set, from which the scientific community can *extract* relevant characteristics and *infer* answers to relevant questions.

This is generally true: model output is transformed into useful knowledge only by a suitable analysis strategy. This strategy is guided by a priori conceptualization, either based on theoretical insight – as in the cases of testing Stommel's theory of the stability of the thermohaline circulation (Sect. 6.1.4) and the determination of the air–sea interaction signature (Sect. 6.2.1) – or on empiricism – as in the case of the circulation statistics of the North Sea (Sect. 6.3.1). The model data need to be projected onto a reduced system[1]. One needs simplified concepts to formulate hypotheses and statistical analyses. An example for such reduction and analysis is the zero-dimensional climate model presented in Sect. 6.2.2, where model data were projected onto an energy balance model, with parameters determined by a statistically optimized fit to the model (and observed) data.

The concepts and statistics do not need to be sophisticated, but attempts to proceed without them will lead to arbitrary conclusions and amount to no more than a fishing expedition[2].

Theoretical and/or empirical analyses and reduction of information are also required to provide meaningful input into models. The reconstruction of past developments or scenarios of future developments require first of all that *a few* potentially important forcing factors have been identified[3]. Numerical experimentation requires well-formulated distinct hypotheses.

Hypothesis testing also identifies the third component where information may need to be reduced, not only in the input and output but in the model itself. For most applications the maxim is that the model should be as realistic as possible, but this is not necessarily true for hypothesis testing. To detect

---

[1] This is also true for observed data. They will reveal "meaning" only after framed within a useful concept. A formal approach is the PIP concept suggested by Hasselmann [57].

[2] ... with the danger that once a fish is caught that its value might not be appreciated.

[3] This is different in those cases of data assimilation where models play the secondary role of augmenting the observational evidence. In these cases all kind of available data are utilized.

signals, to keep track of the causal relations, and to interpret the results often requires that one resorts to less realistic models. The same is true of sensitivity studies, which are often performed with models of intermediate complexity in order to arrive at transparent conclusions. Marotzke [110] and Scott and Marotzke [149] use an idealized ocean general circulation model (rectangular geometry, flat bottom, no wind forcing) to study the sensitivity of the meridional overturning circulation with respect to the magnitude and location of diapycnal mixing.

The issue of reduction of information is also related to the issue of "new" versus "old" models, discussed next.

7.2 New and Old Models

Quasi-realistic models have a finite life time, in contrast to conceptual models. Nobody hesitates to refer to Arrhenius' [3] energy balance model (Sect. 4.2.4), but the idea to use today a climate model that was designed 30 years ago in the mid 1970s would certainly not be applauded. In climate sciences, the life span of a quasi-realistic model version, from its first publication to its eventual demise, is about 5 years. Within these 5 years the model is extended to include more processes, the resolution increased to describe more details, and the code adapted to newer computer technology. The model becomes a new model.

The old models, which just years ago were considered state-of-the-art and adequate to study a wide range of problems, are declared unfit – because more complex new models are now available. When the new version of a model is presented to the scientific community, this often takes the form of first illustrating the deficits of the "old" model and praising the achievements of the "new" model version. However, this "new" model will be "old" in no more than 5 years.

This situation is irritating. If models have a limited lifetime, do their results have a limited life time as well? Are results obtained with models no longer "state-of-the-art" also not "state-of-the art"? Is knowledge generated with quasi-realistic models ephemeral, only valid until the next generation of model version is installed? Certainly not.

First of all, the results of new models do not invalidate *all* the results of old model. Usually, new models change some of the results but mostly they *add* results. Results from "old" models usually remain valid. Many significant results were achieved with the GFDL model system, which for many years was not or only slightly updated[4]. Only rarely are there quantum leaps in the succession of models. For climate models, one example was the replacement of the mixed-layer ocean model by a dynamical 3-dimensional ocean model at around 1990. Another example may be the replacement of the current

---

[4] Another example for a healthy persistent reliance on an "old" modeling system is given by Branstator as discussed in Sect. 6.3.2.

coarse-resolution ocean models (which treat the ocean as made up of mustard rather than of water) by eddy-resolving ocean models. This may change the assessment of the role of the ocean in the climate machinery significantly. Nevertheless, most results obtained with "old" models will remain valid.

Second, more complex models are not necessarily "better" models. They may feature more neutral, potentially positive analogs, but they may also feature fewer of them. More complexity does not imply more positive analogs. Being more complex is not a virtue in itself. Being more complex means that the border of closure has been pushed towards more detailed processes and higher resolution (Sect. 2.1.2). The level at which parameterizations are invoked is changed. More processes are parameterized, more parameters have to be guessed, more assumptions have to be made. Uncertainty increases; it does not decrease.

Third, it must be kept in mind that models are used for a purpose. Again, being more complex is not a virtue for all purposes. Keeping track of causal relations might be more important.

Old models are an important reference to validate new models. Only when the significant results derived from old models (and validated by independent observational evidence) are reproduced by new models, should the new, more detailed results of the new model be considered potentially positive analogs.

We need to value continuity in quasi-realistic modeling, and to be vigilant of the ever-increasing complexity of models, which relies on more and more assumptions encoded somewhere in parameterizations.

7.3 Trustworthiness

Trustworthiness is perhaps the most important issue. It is concerned with the applicability of models and with the potential of models to generate new knowledge. The main points here are that models only describe limited aspects of reality, need to be validated by various means and generate new knowledge about the system only when extrapolated beyond the validated range.

In this book we mostly considered quasi-realistic models, which purport to reproduce a significant part of reality. These models do not describe all aspects of reality. Thus, they are not really "ocean models", but only models which describe certain aspects of the ocean, for instance the large-scale circulation, the water quality or the propagation of sound waves. Principally, models are not models "of" something but models "for" something. They are constructed for a specific purpose, and should in general be applied only for this purpose[5]. The tide models in this book are designed to describe the dynamics of tides on spatial scales larger than a few kilometers in the open ocean and marginal seas. They do not predict the tidal elevations in a seaport or around the

---

[5] It might well happen that a model constructed for one purpose works very well for another, initially not intended purpose as well.

island of Oahu. Climate models are supposed to simulate conditional weather statistics on time scales of days to hundreds and even thousands of years and on spatial scales of several tens or more kilometers, and not to predict the formation of thunderstorms in the next hours and the number of lightning strikes.

There is no way of proving that models give the right answers for the right reasons. Instead the situation is like a judicial process in court, when more and more supporting evidence is gathered for supporting a claim. Complete certainty that the claim is true cannot be achieved, and vigilance is always required. On the other hand, new observations may become available which contradict a model result. Then the model has to be declared inappropriate for certain applications. An example is the stochastically excited 350-year eigen-oscillation in the meridional overturning circulation of the Atlantic Ocean (Sect. 6.1.3), which was a neutral analog. Such a time scale could not be found in proxy data, It was concluded that the real oscillatory modes on these times scales cannot be modeled with prescribed atmospheric boundary values. The neutral analog turned into a negative analog. The ocean model itself is still considered useful, but not for this specific application.

The various ways to validate a model, or to bring supporting evidence before the court, were discussed in Sect. 3.1 and included

- repeated successes in independent forecasts,
- the skill in describing distinctively different configurations, and
- consistency with new (or not yet considered) data sets or with well-established theoretical concepts.

Examples included the success of forecasting in Sect. 5.1, the skill of climate models to simulate both present and paleoclimatic conditions, the consistency of the modeled and reconstructed historical climate during the Late Maunder Minimum in Sect. 6.1.2, and the success of a regional atmospheric model to describe the long-range transport of pollutants in Sect. 5.2.3.

We can usually demonstrate that such models perform more or less skillfully for certain situations which have been well observed. But knowing this, validating models, determining their positive and negative analogs, is not constructive. We only learn something about the model but nothing about reality. This may be interesting for certain people, in particular for mathematicians studying the properties of abstract and often artificial systems. But environmental scientists are supposed to explain the functioning of the real system, to determine its sensitivity to external disturbances, and to predict possible future developments. They must strive for knowledge about the real system.

The constructive use of models is to assume that characteristics of the system, which are represented in the model but which cannot be observed, are valid for the real system as well, that neutral analogs are actually positive analogs , that one can extrapolate beyond the range of past experience. We have provided some examples in Chaps. 5 and 6, in which this bold step is taken: the scenarios in Sect. 5.3, the almost global cooling during the Late

Maunder Minimum (Sect. 6.1.2), the proposition that Stommel's conceptualization governs the stability of the Atlantic overturning in Sect. 6.1.4, or the identification of deep-ocean dissipation as a major part of the tidal energy balance in Sect. 6.3.3.

There are two further aspects about trustworthiness:

- "Good" models do not guarantee good results. The answers given by a quasi-realistic model cannot be better than the questions asked. The principle "rubbish in, rubbish out" applies to quasi-realistic models as well. Useful knowledge is not automatically generated by quasi-realistic models. Instead, modeling strategies need to be carefully designed in a framework based on dynamical hypotheses and empirical evidence, and the output needs to be evaluated in a similar theoretical and statistical framework (cf. Sect. 6.1).
- Models are not impartial umpires. It can happen that the same phenomenon is reproduced in different quasi-realistic models in different ways. This has happened for instance for the cold Late Maunder Minimum episode discussed in Sect. 6.1.2 (See Shindell et al. [152] versus Zorita et al. [196]). Thus, numerical experimentation in itself is not always conclusive in "explaining" phenomena or discriminating between different hypotheses and frameworks.

7.4 Model Builder and Model User

Quasi-realistic computer models are not a common scientific product like a formula or a theory. Quasi-realistic models are usually not constructed by an individual scientist but by a research group, consisting of many individuals with expertise in environmental sciences, applied mathematics and computer technology. The models are executed on computers which require support by hard- and software specialists. Often the models are a product of an institute. The names of models reflect this fact: POM, the Princeton Ocean Model; ECHAM, the European Center HAMburg model; HOPE, the Hamburg Ocean Primitive Equation model. The model builders provide the model to users as executable codes. Often these models are provided to large user communities as *community models*.

There are also other reasons why quasi-realistic models are not a common scientific product:

1. Models are often regarded as a proprietary good, not shared with other scientists.
2. Models are expensive. Only "rich" scientists can afford to build and run models[6].

The users do not necessarily know and understand the inner workings of the code and model. They have to rely on the assertions of the model builders that the model does what it is supposed to do. A quasi-realistic computer model is hence not a scientific product like the theorem of residues or the frequency of the pendulum that has been proven or calculated by generations of mathematicians or physicists.

7.5 Social and Psychological Conditioning

Science is not happening in an isolated space, protected against the traits of human nature. Modelers are humans, who interact with other humans. Some are keen on recognition, others try to avoid conflicts. Science is a social process. The old ideals of "truth speaks to power" have been found to be often unrealistic, in particular in case of environmental sciences. Instead, environmental sciences find themselves in a postnormal stage, where scientific framing, explanation, and perceptions are not only governed by scientific findings but often also by pre-scientific value preferences and even by ideologies. Science takes part in the market of relevant knowledge claims. Scientific success is rewarding not only in terms of recognition but also in terms of job security and promotion. These facts have implications for quasi-realistic modeling.

On the one hand, modelers feel constrained to keep their results within a certain range of what other models produce. Results way off the commonly accepted range are not well received, at least not when presented by file-and-rank scientists. Reinforcing existing concepts with newer, more detailed and complex models is rewarded. Reviewers are pleased to see that their former results and considerations stand the test of time.

On the other hand, modelers also feel tempted to introduce modifications that lead to "spectacular" results. Spectacular results, as long as they are not too speculative, are also well received by such flagships of international publishing as *Science* and *Nature*. Their criterion for publication is "general interest beyond the immediate discipline", though the news must also be able to pass the reviewers, the doorkeepers of conventional wisdom and their own vested interests.

Model output is fairly sensitive to details of the model formulation. Modelers can steer their results in a certain direction by simply changing their codes. Models can be manipulated. In fact, it is a characteristic property of

---

[6] There are only few research centers in first-world countries that are capable of running extensive climate scenarios as those described in the Intergovernmental Panel on Climate Change reports.

models that they can be manipulated. This property is used in numerical experimentation. Manipulation of models is a desired property, but of the experimental set-ups, not of the results. So far, no case of conscious manipulation in the oceanic and atmospheric sciences has been made public, but chances are that they have taken place, and will continue to take place. In other sciences such cases have been discovered and it is unlikely that this specific modeling community is free of similar manipulations.

Many quasi-realistic models are dauntingly complex. Their codes stretch over thousands of lines. Almost certainly these models must contain a number of "errors". Here, "errors" refers to instances when (minor) parts of the code do not function as intended. This sounds worse than it is. Generally these errors will have little consequence on the overall performance of the model. It is believed that these unavoidable errors and inconsistencies will not really affect the model's spectrum of neutral analogs. Thus, this type of error is regarded as a kind of nuisance that one has to live with.

It is, however, a common experience among modelers that model codes and set-ups often also contain more serious errors. Modelers try to eliminate these by carefully repeating simulations and numerical experiments several times under different conditions to see whether the model works as intended. It is only after these tests that the modeler believes that the set-up is free of such significant errors. It is an irritating detail that almost never are cases reported about the detection of such errors *after* publication of the results. No numbers are available but it seems reasonable to assume that at least a few percent of all published results with complex models suffer from serious modeling errors detected too late.

7.6 Final Conclusions

Our main conclusions about the utility of quasi-realistic models are:

- Models in general and quasi-realistic models in particular are major tools to expand our knowledge about environmental systems.
- Models of environmental systems are neither true or untrue, nor valid or invalid per se; rather they are adequate or inadequate, useful or not useful to answer specific questions about the system. They describe larger or smaller parts of reality, but never all of reality.
- The validation of models by comparison with data is an important prerequisite for their application. However, the validation does only provide new insight about the model. New knowledge about the studied system is gained by applying models to new situations outside the validated range with all the risks that such an extrapolation entails.
- Models are a tool only. They do not divulge new knowledge by themselves. The act of generating new knowledge still depends on the modeler, on his or her skill in setting up the simulation or experiment, in asking the right questions, and in applying the right methods to analyze the model output.

Appendices

In the following appendices we describe in some detail the technical background for four important aspects of modeling:

A: Foundations of the dynamical equations used in atmospheric and oceanic modeling,
B: Aspects of the numerical formulation of dynamical equations,
C: Aspects of statistical terminology and concepts, and
D: Concepts of data assimilation.

The presentation is rather compact and makes use of vector and tensor notation.

A

Fluid Dynamics

In this appendix we describe in some detail the dynamical laws that govern fluid systems such as the ocean or the atmosphere. These laws are the subject of *fluid dynamics*. Fluid dynamics combines three types of laws: the conservation laws, or more generally, the balance equations for mass, momentum and energy; the thermodynamic laws that govern the properties of individual fluid parcels; and the phenomenological flux laws. These laws are well established and proven experimentally. They are discussed first, in Sect. A.1 through Sect. A.4. Next we describe in Sect. A.5 the additions and modifications to the balance equations when radiation, phase transitions and photochemical reactions are included. These sections provide the basic physics underlying the dynamics of the atmosphere and ocean.

The next sections cover some of the approximations when one applies these basic laws to geophysical flows, i.e., to large-scale flows in the atmosphere and ocean. The Reynolds decomposition is explicitly treated in Sect. A.6. The standard parameterizations of interior eddy fluxes in terms of eddy diffusion coefficients and of boundary eddy fluxes in terms of drag coefficients and "Ekman suction" are given in Sects. A.7 and A.8. The anelastic and shallow water approximations to the basic fluid dynamical equations are discussed in Sect. A.9 and examples of different representations are given in Sect. A.10.

A.1 The Balance Equations

A.1.1 Mass Balances

Fluids like seawater and air consist of many components. Here we assume, as is often done, that they consist of just two components: water and sea salt in the case of sea water, and dry air and water vapor in the case of air. This assumption is not essential and can easily be relaxed, as is demonstrated in Sect. A.5. Each of the two components is characterized by its density ρ_i $(i = 1, 2)$ which is the mass of component i per unit volume. The densities depend on position $\boldsymbol{x}$ and time t. The mass of component i in a fixed volume V changes according to

$$\frac{d}{dt} \iiint_V d^3x \, \rho_i(\boldsymbol{x}, t) = -\iint_A d^2x \, \boldsymbol{n} \cdot \boldsymbol{J}_i(\boldsymbol{x}, t) + \iiint_V d^3x \, S_i(\boldsymbol{x}, t) \qquad (A.1)$$

for $i = 1, 2$. Here $\boldsymbol{J}_i$ is the mass flux into or out of the volume V through the bounding surface A with outward normal vector $\boldsymbol{n}$. If the volume is fixed this flux is given by

$$\boldsymbol{J}_i = \rho \, \boldsymbol{u}_i \qquad (A.2)$$

where $\boldsymbol{u}_i(\boldsymbol{x}, t)$ is the velocity of component i. $S_i(\boldsymbol{x}, t)$ is a source or sink term that describes the generation or destruction of component i within the volume V, e.g., by chemical reactions. The balance equation (A.1) is self-evident. To change the mass of a component within a volume one has either to transfer mass across the bounding surface or to generate or annihilate it inside the volume. More formally, the balance equations define fluxes and sources. Since the volume V is arbitrary in (A.1) Gauss's theorem implies the differential form

$$\partial_t \rho_i + \boldsymbol{\nabla} \cdot \boldsymbol{J}_i = S_i \qquad (A.3)$$

If $S_i = 0$ then the mass of component i is said to be (globally) conserved. The mass of component i does not change in a volume moving with the velocity $\boldsymbol{u}_i$.

Next introduce the total mass density[1] $\rho := \rho_1 + \rho_2$ and the fluid velocity $\boldsymbol{u} := (\rho_1 \boldsymbol{u}_1 + \rho_2 \boldsymbol{u}_2)/(\rho_1 + \rho_2)$, which is the velocity of the center of mass of the two components, often called the barycentric velocity. The fluxes now become

$$\boldsymbol{J}_i = \rho_i \boldsymbol{u} + \boldsymbol{I}_i \qquad (A.4)$$

where $\boldsymbol{I}_i = \rho_i(\boldsymbol{u}_i - \boldsymbol{u})$. The first term in (A.4), $\rho_i \boldsymbol{u}$, is called the advective flux and $\boldsymbol{I}_i$ the diffusive flux. The diffusive flux comes about because the component i does not move with the fluid velocity $\boldsymbol{u}$ but with its own velocity $\boldsymbol{u}_i$. If there are no sources or sinks the balance equation for the total mass becomes

$$\partial_t \rho + \boldsymbol{\nabla} \cdot (\rho \, \boldsymbol{u}) = 0 \qquad (A.5)$$

It is called the *continuity equation*. There is no diffusive flux for the total mass since $\boldsymbol{I}_1 + \boldsymbol{I}_2 = 0$. The continuity equation for either one of the components, say component 2, is

$$\partial_t \rho_2 + \boldsymbol{\nabla} \cdot (\rho_2 \boldsymbol{u} + \boldsymbol{I}_2) = 0 \qquad (A.6)$$

Instead of the density ρ_2 one often uses the concentration $c_2 = \rho_2/\rho$, which is the mass of component 2 per unit total mass. Its equation is

$$\rho \frac{D}{Dt} c_2 = -\boldsymbol{\nabla} \cdot \boldsymbol{I}_2 \qquad (A.7)$$

where $D/Dt = \partial_t + \boldsymbol{u} \cdot \boldsymbol{\nabla}$ is the advective or material derivative. Usually, c_2 is the salinity for sea water and the specific humidity for air.

---

[1] The notation $a := b$ is to be read such that a is defined by the expression b.

A.1.2 Momentum Balance

Newton's second law states $m\ddot{\boldsymbol{x}} = \boldsymbol{F}$. The acceleration $\ddot{\boldsymbol{x}}$ of a particle of mass m is given by the force $\boldsymbol{F}$ acting on the particle. This is a useful law only if the force can be prescribed independently. For the gravitational force this independent prescription is $\boldsymbol{F} = -m\,\boldsymbol{\nabla}\phi_g$, where ϕ_g is the gravitational potential. The generalization of Newton's law to continuous fluid systems takes the form

$$\rho\frac{D}{Dt}\boldsymbol{u} = -\boldsymbol{\nabla}p + \boldsymbol{\nabla}\cdot\boldsymbol{\sigma} - \rho\,\boldsymbol{\nabla}\phi_g \qquad (A.8)$$

Two new forces appear: the pressure force given by the gradient of the pressure p and the frictional forces given by the divergence of the viscous stress tensor $\boldsymbol{\sigma}$. The two new forces describe the effect that other fluid parcels exert on the fluid parcel under consideration. The gravitational potential is due to the gravity field of the earth, generally characterized by the gravitational acceleration g. In tidal problems one has to include the tidal potential due to the gravity fields of the moon and sun.

Since the pressure and viscous forces are surface forces the momentum equation (A.8) can be cast into the standard form of a balance equation[2]

$$\partial_t(\rho\boldsymbol{u}) + \boldsymbol{\nabla}\cdot(\rho\,\boldsymbol{uu} + p\,\boldsymbol{I} - \boldsymbol{\sigma}) = -\rho\,\boldsymbol{\nabla}\phi_g \qquad (A.9)$$

where $\boldsymbol{I}$ is the unit tensor. The interpretation then is that $\rho\boldsymbol{u}$ is the momentum density and that the momentum flux consists of the advective contribution $\rho\boldsymbol{uu}$, a contribution due to the pressure, and a contribution given by the viscous stress tensor. The viscous stress tensor describes momentum diffusion.

In a rotating system such as the earth one must add the Coriolis and centrifugal forces to the right hand side of the momentum balance. The Coriolis force is $\boldsymbol{F}_c = -\rho\,2\,\boldsymbol{\Omega}\times\boldsymbol{u}$ where $\boldsymbol{\Omega}$ is the rate of rotation. The centrifugal force can be derived from a potential ϕ_c which is generally combined with the gravitational potential ϕ_g to form the geopotential ϕ.

A.1.3 Energy Balance

Energy comes in various forms. The kinetic energy per unit mass is given by

$$e_k = \frac{1}{2}\boldsymbol{u}\cdot\boldsymbol{u} \qquad (A.10)$$

and the potential energy per unit mass by the geopotential

$$e_p = \phi \qquad (A.11)$$

---

[2] In this appendix we use the tensor notation commonly adopted by theoretical physics. For those not familiar with this notation, we explain it in footnotes: **uu** is the product of two vectors (tensor of 1st order), the result of which is a matrix (tensor of 2nd order).

Together they form the mechanical energy. The first law of thermodynamics states that there exists another form of energy, the internal energy e_i, and that mechanical and internal energy can be converted into each other such that the total energy is conserved. For a continuous fluid system this first law again takes the form of a balance equation

$$\partial_t(\rho\, e_i) + \nabla \cdot (\rho\, e_i\, \boldsymbol{u} + \tilde{\boldsymbol{q}}) = Q \tag{A.12}$$

Here $\tilde{\boldsymbol{q}}$ is the heat flux due to conduction or diffusion of heat. The source term Q is given by[3] $Q = (-p\,\boldsymbol{I} + h\boldsymbol{\sigma}) : \boldsymbol{D}$ where $\boldsymbol{D}$ is the rate of deformation or rate of strain tensor with components $D_{ij} = \frac{1}{2}(\partial_i u_j + \partial_j u_i)$. The source term Q describes the reversible exchange of mechanical and internal energy by pressure forces and the irreversible conversion of mechanical energy into internal energy by viscous forces. Equations for the kinetic, potential and total energy can be derived from (A.12) with the help of the momentum balance.

The balance equations for mass, momentum and energy all have the same structure. The amount of mass, momentum or energy within a volume changes by fluxes across the bounding surface and by sources or sinks within the volume. The balance equations are prognostic equations for the mass densities ρ_1 and ρ_2 (or ρ and ρ_2), the momentum density $\rho\,\boldsymbol{u}$, and the internal energy density $\rho\, e_i$. To calculate the evolution of these quantities one needs to specify the geopotential ϕ and the earth's rotation rate $\boldsymbol{\Omega}$, the diffusive fluxes $\boldsymbol{I}_2$, $\boldsymbol{\sigma}$ and $\tilde{\boldsymbol{q}}$, and the pressure p. In addition one needs to provide boundary and initial conditions. Let us first turn to the specification of the pressure which requires us to look at the thermodynamics of fluids.

A.2 Thermodynamic Specification

Fluid dynamics assumes local thermodynamic equilibrium. Each fluid parcel is in thermodynamic equilibrium, although the whole system is not. Each fluid parcel can thus be described by thermodynamic variables such as the density ρ, the pressure p, and the temperature T. Gibbs' phase rule states that the number of variables required to completely specify the thermodynamic state of a fluid particle is given by

$$f = \chi - \varphi + 2 \tag{A.13}$$

where χ is the number of components and φ the number of phases. For a one-component, one-phase system $f = 2$. In this case one usually uses p and T as the independent variables. Other variables, such as the density ρ, are then determined by p and T. The relation $\rho = \rho(p, T)$ is called the equation of state. If two phases coexist in a one-component system then $f = 1$. If one chooses p as the independent variable then relations $T_b = T_b(p)$ or $T_f = T_f(p)$ determine the boiling or freezing temperatures.

---

[3] $\boldsymbol{a} : \boldsymbol{b}$ is the product of two second order tensors that results in a scalar.

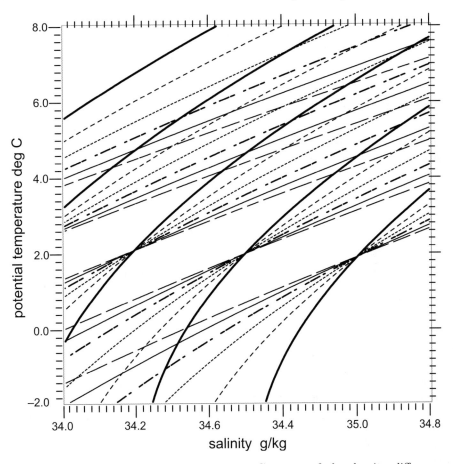

Fig. A.1. Equation of state for sea water. Contours of the density difference $\rho(p,T,S) - \rho(p,2°C,34,5\,\text{psu})$ are shown in the T-S plane for different values of pressure corresponding to depths of 0 m (*thick solid*) to 5 km (*long dashed*) in 1 km intervals. The contour interval is $0.25\,\text{kg/m}^3$. The equation of state is nonlinear. The contours (isopycnals) are *curved* and their slope turns with pressure. Courtesy of Ernst Maier-Reimer

We assumed sea water or air to be a two-component, one-phase system. Then $f = 3$. One could choose ρ, c_2 and e_i as the three independent variables, where c_2 is the salinity S for sea water or the specific humidity q for air. These variables then determine the pressure, $p = p(\rho, c_2, e_i)$. Other variables such as the temperature T are also determined by (ρ, c_2, e_i). However, fluid dynamicists prefer to use (p, T, c_2) as the independent variables. They thus have to transform the prognostic equations for ρ and e_i into prognostic equations for p and T. This is straightforward. One just has to differentiate the relations $p = p(\rho, c_2, e_i)$ and $T = T(\rho, c_2, e_i)$ and substitute the prognostic equations for ρ, c_2 and e_i. The result is

$$\frac{D}{Dt}T = -\rho c^2 \Gamma \, \boldsymbol{\nabla} \cdot \boldsymbol{u} - \frac{1}{\rho c_V}(\boldsymbol{\nabla} \cdot \tilde{\boldsymbol{q}} - \boldsymbol{\sigma} : \boldsymbol{D}) \qquad (A.14)$$

$$\frac{D}{Dt}p = -\rho c^2 \boldsymbol{\nabla} \cdot \boldsymbol{u} - \frac{\rho c^2 \Gamma}{T}(\boldsymbol{\nabla} \cdot \tilde{\boldsymbol{q}} - \boldsymbol{\sigma} : \boldsymbol{D}) \qquad (A.15)$$

New thermodynamic coefficients enter these equations: the speed of sound c, the adiabatic temperature gradient Γ, and the specific heat at constant volume c_V. All these coefficients depend on (p, T, c_2). Also, the density ρ now becomes a dependent variable. Thus the prognostic equation for p, T, c_2 and $\boldsymbol{u}$ have to be augmented by the relations

$$\begin{aligned} \rho &= \rho(p, T, c_2) \\ c &= c(p, T, c_2) \\ \Gamma &= \Gamma(p, T, c_2) \\ c_V &= c_V(p, T, c_2) \end{aligned} \qquad (A.16)$$

These relations are diagnostic relations. They describe properties of the fluid. They have been measured and are tabulated for sea water and air (see Fig. A.1).

A.3 The Phenomenological Flux Laws

The diffusive fluxes $\boldsymbol{I}_2$, $\boldsymbol{q}$ and $\boldsymbol{\sigma}$ represent the effect that salt (or water vapor), heat and momentum are not only transported with the fluid velocity but also by the mechanisms of molecular diffusion, molecular heat conduction and molecular friction or viscosity. There exists a well-developed theory, the theory of non-equilibrium or irreversible thermodynamic processes, that deals with the determination of such molecular fluxes. The theory involves three steps.

1. It is shown in thermodynamics that a system in an external potential is in thermodynamic equilibrium if

$$\begin{aligned} T &= \text{constant} \\ \mu_i + \phi &= \text{constant} \quad i = 1, 2 \\ \boldsymbol{D} &= 0 \end{aligned} \qquad (A.17)$$

where μ_i is the chemical potential of component i. First, the temperature must be constant. The second condition implies that the chemical potential difference $\Delta\mu = \mu_2 - \mu_1$ must be constant. The third condition, the vanishing of the rate of strain tensor, implies that the system can only move with a uniform translation and rotation. If an external field is present, thermodynamic equilibrium does not imply that the concentrations and the pressure are constant. Instead we have the hydrostatic balance

$$\nabla p = -\rho\,\nabla\phi \tag{A.18}$$

and

$$\nabla c_2 = \gamma\,\nabla p \tag{A.19}$$

where

$$\gamma = -\frac{(\partial\Delta\mu/\partial p)_{T,c_2}}{(\partial\Delta\mu/\partial c_2)_{p,T}} \tag{A.20}$$

describes the equilibrium concentration gradient.

2. It can also be shown that the rate of entropy production $\dot\eta$ in a fluid is given by

$$\dot\eta = \mathbf{I}_2 \cdot \left(-\frac{1}{T}\nabla\,\Delta\mu|_T\right) + \mathbf{q}\cdot\nabla\left(\frac{1}{T}\right) + \boldsymbol{\sigma} : \left(\frac{1}{T}\mathbf{D}\right) \tag{A.21}$$

It is the sum of three terms. All terms are the product of a flux and a "force". In thermodynamic equilibrium all the "forces" are zero and all the fluxes are zero.

3. The basic assumption of the theory of irreversible processes is that there exists a linear relationship between the fluxes and the forces. If the fluid is isotropic and if one expresses the gradient of the chemical potential difference at constant T, $\nabla\Delta\mu|_T$, in terms of the concentration and pressure gradients these linear relationships take the form

$$\begin{aligned}
\mathbf{I}_2 &= -\rho[\kappa_S(\nabla S - \gamma\,\nabla p) + \kappa_{ST}\nabla T]\\
\mathbf{q} &= -\rho[\kappa_T\nabla T + \kappa_{TS}(\nabla S - \gamma\,\nabla p)]\\
\boldsymbol{\sigma} &= 2\,\rho\,\nu\,\mathbf{S} + 3\,\rho\,\nu'\,\mathbf{N}
\end{aligned} \tag{A.22}$$

given here for the case that $c_2 = S$. These relations are called the phenomenological flux laws. Here $\mathbf{D} = \mathbf{S} + \mathbf{N}$ where $\mathbf{S}$ is the rate of shear deformation tensor, which has trace zero, and $\mathbf{N}$ is the rate of normal deformation tensor[4].

The phenomenological flux laws state that the salt flux $\mathbf{I}_2$ is proportional to the deviation of the salinity gradient from its equilibrium value (A.19). The factor of proportionality is the salt diffusion coefficient κ_S. Salt diffusion is also caused by temperature gradients. The coefficient is the thermo-diffusion coefficient κ_{ST}. Similarly, heat diffusion or conduction is driven by temperature gradients with the factor of proportionality being the thermal conduction coefficient κ_T and by deviations of the salinity gradient from its equilibrium value. The coefficient κ_{TS} for the latter process is not independent but related to κ_{ST} by

$$\kappa_{TS} = T\left(\frac{\partial\,\Delta\mu}{\partial S}\right)_{p,T}\kappa_{ST} \tag{A.23}$$

---

[4] Note that the heat flux $\mathbf{q}$ in (A.22) and (A.21) differs from the heat flux $\tilde{\mathbf{q}}$ in (A.12), (A.14) and (A.15). The two are related by $\tilde{\mathbf{q}} = \mathbf{q} + \mathbf{I}_2\Delta h$ where Δh is the partial enthalpy difference.

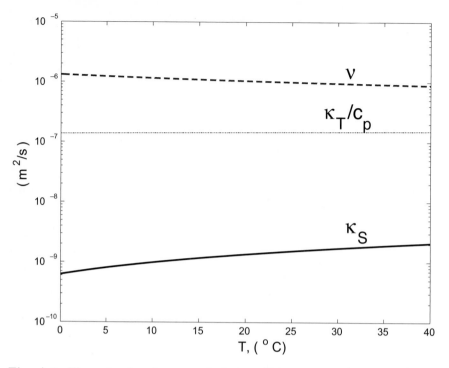

Fig. A.2. The molecular phenomenological coefficients κ_S, κ_T/c_p, and ν for sea water as a function of temperature T for $S = 35\,\mathrm{psu}$ (32.9 psu for κ_S) and $p = 0$

This relation is a consequence of the "Onsager relations" which state that the linear relation between fluxes and forces is symmetric. The viscous or frictional stresses are driven by the rate of shear deformation with shear viscosity ν and by the rate of normal deformation with expansion viscosity ν'. The flux laws (A.22) thus contain five independent phenomenological or molecular diffusion coefficients, namely κ_S, κ_T, κ_{ST}, ν and ν'. The second law of thermodynamics requires $\dot{\eta} \geq 0$ in (A.21). This implies $\kappa_S \geq 0$, $\kappa_T \geq 0$, $\nu \geq 0$ and $\nu' \geq 0$. The coefficient κ_{ST} can have arbitrary sign. The coefficients are properties of the fluid. They depend on (p, T, S). Except for ν' they have been measured and are tabulated. For sea water (see Fig. A.2) one finds $\kappa_S < \kappa_T/c_p < \nu$ where c_p is the specific heat at constant pressure. Thus momentum diffuses at a faster rate than heat, and heat faster than salt.

The phenomenological flux laws are often written in different but equivalent forms and in terms of different but equivalent phenomenological coefficients. The flux laws for air have the same structure.

A.4 Boundary Conditions

The balance equations are partial differential equations in space and time. Their solution requires the specification of boundary and initial conditions. Here we consider the boundary conditions.

A boundary is a surface $G(\boldsymbol{x}, t) = 0$. It has normal vector $\boldsymbol{n} = \boldsymbol{\nabla} G / |\boldsymbol{\nabla} G|$ and moves with velocity $\boldsymbol{v} = -\partial_t G \, \boldsymbol{n} / |\boldsymbol{\nabla} G|$. The fluxes of mass, salt, momentum and total energy through such a moving boundary are given by

$$
\begin{aligned}
F_\rho &= \rho(\boldsymbol{u} - \boldsymbol{v}) \cdot \boldsymbol{n} \\
F_S &= [\rho \, S(\boldsymbol{u} - \boldsymbol{v}) + \boldsymbol{I}_2] \cdot \boldsymbol{n} \\
\mathbf{F}_{\boldsymbol{u}} &= [\rho \, \boldsymbol{u}(\boldsymbol{u} - \boldsymbol{v}) - \boldsymbol{\Pi}] \cdot \boldsymbol{n} \\
F_{e_t} &= [\rho \, e_t(\boldsymbol{u} - \boldsymbol{v}) + \boldsymbol{q} - \boldsymbol{u}\boldsymbol{\Pi}] \cdot \boldsymbol{n}
\end{aligned}
\tag{A.24}
$$

where $\boldsymbol{\Pi} = -p\,\boldsymbol{I} + \boldsymbol{\sigma}$ is the stress tensor.

Consider first a completely isolated system that does not exchange any mass, salt, momentum and energy with its surroundings. In this case all the fluxes are zero, which implies

$$
\begin{aligned}
\boldsymbol{u} \cdot \boldsymbol{n} &= \boldsymbol{v} \cdot \boldsymbol{n} \\
\boldsymbol{I}_2 \cdot \boldsymbol{n} &= 0 \\
\boldsymbol{\Pi} \cdot \boldsymbol{n} &= 0 \\
\boldsymbol{q} \cdot \boldsymbol{n} &= 0
\end{aligned}
\tag{A.25}
$$

The first of the equations states that the surface moves with the fluid velocity. It is hence made up of the same fluid particles at all times. The surface is a material surface.

However, no system is truly isolated. It exchanges mass, momentum and energy with its surroundings. It is open, and mechanically and thermally coupled to its surroundings. In this case one can develop phenomenological flux laws similar to the ones derived in the previous section. This leads to expressions of the form (A.22) but with the gradients $\boldsymbol{\nabla} S, \boldsymbol{\nabla} p, \boldsymbol{\nabla} T, \boldsymbol{\nabla} \boldsymbol{u}$ on the right hand side replaced by the jumps $\Delta S, \Delta p, \Delta T, \Delta \boldsymbol{u}$ across the boundary. These laws are similar to Ohm's law for a resistor that states that the electrical current is proportional to the difference of the electrical potential. In fluid dynamics one considers, however, the limit that the surfaces have no resistance. They are short-circuited. In this limit one obtains the following conditions at the boundaries:

1. the fluxes of mass, salt, momentum and total energy must be continuous across the boundary

$$
\begin{aligned}
F_\rho^I &= F_\rho^{II} \\
F_S^I &= F_S^{II} \\
\mathbf{F}_{\boldsymbol{u}}^I &= \mathbf{F}_{\boldsymbol{u}}^{II} \\
F_{e_t}^I &= F_{e_t}^{II}
\end{aligned}
\tag{A.26}
$$

2. the tangential velocity components, temperature and chemical potentials must be continuous across the boundary

$$\boldsymbol{u}^I \times \boldsymbol{n} = \boldsymbol{u}^{II} \times \boldsymbol{n}$$
$$T^I = T^{II} \qquad\qquad (A.27)$$
$$\mu_i^I = \mu_i^{II} \quad i = 1, 2$$

Here the superscripts I and II denote the two media adjacent to the boundary. The first set of conditions is required since mass, salt, momentum and total energy cannot accumulate in an interface of zero volume. The second set is a consequence of the no-resistance assumption. Note that only the tangential velocity is required to be continuous. The normal component can have a discontinuity if there is a mass flux across the interface. In this case the surface is not a material surface. The continuity of the chemical potentials implies that the specific humidity equals the saturation specific humidity $q_S = q_S(p, T, S)$. These conditions also tacitly assume that the fluid does not detach from its boundary.

The main point to note about the boundary conditions is that they are conditions of continuity. It is only in limiting cases that they become prescriptions. Such limiting cases are:

- Rigid boundaries. The surrounding medium is often assumed to be rigid, *i.e.*, to have zero elasticity. Such a rigid surrounding will absorb any momentum flux without moving. The continuity conditions for momentum flux and velocity thus become the prescription $\boldsymbol{u} = 0$ at the rigid boundary.
- Infinite heat capacity. If the surrounding medium has an infinite heat capacity it stays at the same temperature T^* which is not affected by the exchange of heat with the system under consideration. The continuity conditions for temperature and heat flux thus become the prescription $T = T^*$ at the boundary.

A.5 A Closer Look at the Balance Equations

As one looks more closely at fluids like the ocean or atmosphere, more and more components and processes come into play. Water in the atmosphere exists in all its three phases, as water vapor and as liquid drops and ice crystals in clouds; and one has to consider exchanges between these different phases. In addition to mechanical and internal energy there is radiative energy; and one has to consider exchanges between these different energy forms when radiation is absorbed or emitted by matter. Absorption and emission of radiation often involves chemical reaction; and one has to consider photochemical reactions. All these additional components and processes can be accounted for by introducing additional balance equations. How this is done in principle will be demonstrated in this section for the cases of cloud formation, radiation and photochemical reactions.

A.5.1 Cloud Formation

The primitive equation given in Sect. 4.2.3 treats the atmosphere as a two-component system, consisting of dry air and water vapor. The equations contain two mass balances, one for the total mass and one for the water vapor. A specified cloud distribution can be included in the temperature equation to account for the amount of radiation that is absorbed or emitted by clouds. The cloud distribution is, however, not calculated prognostically. To do so one has to consider the physical mechanisms by which clouds are formed and broken up. Clouds are formed when water vapor condenses into cloud drops or deposits into ice crystals. They are broken up when the cloud drops or ice crystals evaporate or sublimate or precipitate as rain or snow. To account for these phase changes and processes one has first to distinguish between the three phases of water and formulate separate mass balance equations for them. These mass balances must include the rates by which mass is transferred from one phase to another. The three mass balances thus have the form

$$
\rho \frac{D}{Dt} c_v = \ldots - S_c + S_e - S_d + S_s
$$
$$
\rho \frac{D}{Dt} c_l = \ldots + S_c - S_e + S_m - S_f \qquad \text{(A.28)}
$$
$$
\rho \frac{D}{Dt} c_s = \ldots + S_d - S_s - S_m + S_f
$$

with the concentrations c_v, c_l and c_s of water in its vapor (v), liquid (l) and solid (s) phase, and the rates of condensation S_c, evaporation S_e, deposition S_d, sublimation S_s, melting S_m and freezing S_f. The dots account for other processes that affect the mass balances, such as diffusion or precipitation.

The phase changes also require or release latent heat which must be accounted for in the internal energy or temperature equation

$$
c_p \frac{D}{Dt} T = \ldots - L_e(S_e - S_c) - L_m(S_m - S_f) - L_s(S_s - S_d) \qquad \text{(A.29)}
$$

where L_e, L_m and L_s are the latent specific heats of evaporation, melting and sublimation. Also the equation of state and other thermodynamic relations need to be modified.

The major task is, of course, to specify the various rates in (A.28, A.29) at which these phase changes occur. The physical processes that determine these rates are discussed in detail in text books on the microphysics of clouds (e.g., [135]). For condensation two of the processes are "nucleation" and "diffusional growth".

Nucleation. Inelastic collisions between water vapor molecules in a supersaturated atmosphere lead to aggregates. These aggregates will survive and form the nucleus of a cloud drop if the surface tension work required to increase the surface is less then the latent heat released by condensation. Since the surface tension work is proportional to r^2 and the latent heat release proportional to r^3 there is a critical radius beyond which the aggregate will survive. This critical radius can be estimated from equilibrium thermodynamics. This critical radius becomes smaller when nucleation occurs at hygroscopic aerosol particles, since the resulting solution has a lower saturation pressure than pure water. Once the critical radius is determined the formation rate of nuclei can be determined from probabilistic calculations.

Diffusional growth. Cloud drops grow by diffusion of water vapor towards the drop as long as the surrounding water vapor pressure exceeds the saturation pressure of the drop. The latent heat released by condensation diffuses away from the drop. Cloud drops evaporate when the saturation pressure of the drop is larger than the surrounding water vapor pressure. Then water vapor diffuses away from the drop and heat towards the drop. For a single drop the growth/decay rate of the drop radius can be estimated from the diffusion equations for water vapor and heat. The most important result of this calculation is that the change of drop radius and hence the condensation rate depend on the drop radius. Thus, the condensation rate S_c in the balance equations (A.29) depends not only on the concentration $c_l(\boldsymbol{x},t)$ of liquid water in clouds but also on the distribution of this liquid water over drops of different radii. A balance equation for the "size spectrum" $c_l(\boldsymbol{x},r,t)$ needs to be developed.

A.5.2 Radiation

In Sect. A.1.3 we considered the energy equation. We only considered the mechanical and internal energies. Electromagnetic radiation is also a form of energy. Radiative energy is converted to internal energy when radiation is absorbed by matter. Internal energy is converted to radiative energy when matter emits radiation. These absorption and emission processes are central to climate modeling since it is the incoming radiation from the sun that drives the ocean–atmosphere system.

Electromagnetic radiation also obeys a balance equation called the radiation balance equation. For natural unpolarized radiation it has the form

$$\partial_t I(\boldsymbol{x},t) + \boldsymbol{\nabla} \cdot \boldsymbol{F}(\boldsymbol{x},t) = S(\boldsymbol{x},t) \tag{A.30}$$

where I is the energy density of the radiation, $\boldsymbol{F}$ the radiative flux, and S a source term describing the absorption and emission of radiation by matter. The term S must be added with the opposite sign to the internal energy equation (A.12). Since all the processes considered in climate modeling evolve at a speed much smaller than the speed of light the term $\partial_t I$ in (A.30) can safely be neglected and S be equated to the divergence of the radiative flux $\boldsymbol{F}$. We kept the time derivative or storage term to demonstrate that radiation obeys a balance equation of the same form as any other extensive quantity.

Closer inspection of radiation processes reveals that they depend strongly on the wavelength λ of the radiation. One must distinguish between short-wave or solar radiation and long-wave or terrestrial radiation. Radiative processes also depend on the direction of the radiation as *e.g.* seen in Fresnel's laws of reflection and refraction. One therefore introduces the wavenumber vector $\boldsymbol{k}$, whose magnitude is given by the wavelength λ of the radiation, $|\boldsymbol{k}| = 2\pi/\lambda$, and whose direction is given by the direction of the radiation, and spectral densities which describe the distribution of the radiative energy, flux and source function with respect to wavenumber. The balance equation for these spectral densities takes the form

$$\partial_t I(\boldsymbol{k}, \boldsymbol{x}, t) + \boldsymbol{\nabla} \cdot \boldsymbol{F}(\boldsymbol{k}, \boldsymbol{x}, t) + \boldsymbol{\nabla}_k \cdot \boldsymbol{R}(\boldsymbol{k}, \boldsymbol{x}, t) = S(\boldsymbol{k}, \boldsymbol{x}, t) \qquad (A.31)$$

where $\boldsymbol{\nabla}_k$ is the gradient in wavenumber space and $\boldsymbol{R}$ a flux in wavenumber space which arises when waves are refracted in a medium with varying index of refraction. Equation (A.31) is a balance equation in physical and wavenumber space. It forms the general framework to describe radiative processes in the atmosphere and ocean.

Again, one can safely neglect the storage term, and in most circumstances also the refractive term and the horizontal flux divergence. In this case one writes $\boldsymbol{F} = F\,\boldsymbol{n}$ where $\boldsymbol{n} = \boldsymbol{k}/k$ is the direction of the radiation and introduces the wavelength λ, the zenith angle $\theta (0 \leq \theta \leq \pi)$, and the azimuthal angle $\varphi (0 \leq \varphi < 2\pi)$ of the radiation, instead of $\boldsymbol{k}$. The radiation balance equation for $F(\lambda, \theta, \varphi, z, \boldsymbol{x}_h, t)$ then reduces to

$$\frac{\partial F}{\partial z} = k^a F \, \sec\theta \qquad (A.32)$$

$$+ k^s \sec\theta \left[F - \int d\theta' \int d\varphi' \gamma(\theta', \varphi', \theta, \varphi) \sin\theta' F(\theta', \varphi') \right] - E$$

which is called the Schwarzschild equation. The first term describes the absorption of radiation. The absorption coefficient k^a describes the absorption per unit volume and has dimensions L^{-1}. The geometric factor $\sec\theta$ takes into account that the optical path through a layer of thickness dz increases by a factor $\sec\theta$ for radiation incident at angle θ. The second term describes the scattering of radiation with scattering coefficient k^s. It consist of two parts. The first part describes the loss due to scattering out of the direction (θ, φ). The second part describes the gain due to the scattering out of directions (θ', φ') into direction (θ, φ), with γ being the normalized scattering cross section. The third term, E, describes the emission. For matter in thermal equilibrium Kirchoff's law states that the emissivity equals the absorptivity or

$$E = k^a F^*(\lambda) \sec\theta \qquad (A.33)$$

where F^* is the isotropic black body radiation given by

$$F^* = \frac{2\pi c^2 h}{\lambda^5} [\exp(c\,h/\lambda\,k\,T) - 1]^{-1} \qquad (A.34)$$

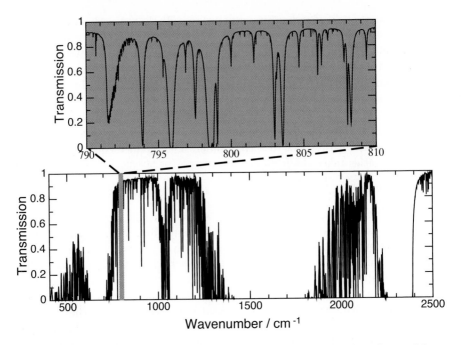

Fig. A.3. Calculated transmission coefficient of a standard atmosphere without clouds as a function of wavenumber using tabulated values for the absorption coefficients. Only the absorber H_2O, CO_2, O_3, CH_4 and N_2O are included. Courtesy of Heinz-Dieter Hollweg and Stephan Bakan

Here c is the speed of light, h Planck's constant, and k Boltzmann's constant.

As discussed in Chaps. 1 and 4 there are many radiatively active constituents in the atmosphere. Among the most important ones are water vapor, carbon dioxide and ozone for long wave radiation and clouds, water vapor and ozone for short wave radiation. They absorb, scatter and emit radiation at different wavelengths and rates. If there are N radiatively active components then

$$k^{a,s} = \sum_{i=1}^{N} c_i k_i^{a,s} \qquad (A.35)$$

where c_i is the concentration of the ith constituent and $k_i^{a,s}$ the absorption or scattering coefficient of the ith constituent. The individual absorption and scattering coefficients are properties of the constituent and depend on the wavelength λ, temperature T and pressure p, $k_i^{a,s} = k_i^{a,s}(\lambda, T, p)$. They have been determined by measurements and quantum-mechanical calculations for most radiatively active components of the atmosphere. The results are collected and updated in large data bases. The total absorption and scattering coefficients depend additionally on the concentrations. Figure A.3 shows a calculation of the transmission coefficient of the atmosphere using individual absorption coefficients and concentrations. The absorption and resulting transmission coefficients consist of lines and bands in wavenumber space. This finestructure causes problems when wavenumber averages of nonlinear relationships are employed in measurements or applications.

In summary, the inclusion of radiative processes requires first the addition of a radiation balance equation, such as (A.32). Secondly, the internal energy or temperature equation needs to be modified to account for the exchange of radiative and internal energy. Thirdly, the radiation balance equation contains the concentrations or densities of the radiatively active constituents, through (A.35); one thus has to add balance equations for these constituents. In the case of water vapor and clouds this requires the inclusion of phase transitions; in the case of ozone one has to include photochemical reactions, which are discussed next.

A.5.3 Photochemical Reactions

The absorption and emission of radiation often involves chemical reactions. Here we introduce the reaction rate of such photochemical reactions and consider as a particular example the creation and destruction of ozone in the oxygen cycle.

Ozone is radiatively active substance, i.e., it interacts with photons. We denote a photon by its energy $h\nu$ where h is Planck's constant and ν its frequency.

We follow chemical notation and denote by $[M]$ the number of molecules M per unit volume. The number of photons of energy $h\nu$ is denoted by $[h\nu]$.

Ozone is created by photodissociation of oxygen. This process occurs in two steps. In the first reaction $O_2 + h\nu_1 \rightarrow O + O$, an ultraviolet photon $h\nu_1$ dissociates an oxygen molecule O_2 into two oxygen atoms. The photon has to have an energy larger than the binding energy $h\nu_1^e$ of O_2. In the second reaction $O + O_2 + M \rightarrow O_3 + M$, one oxygen atom O reacts with one oxygen molecule O_2 to form one ozone molecule O_3. This reaction requires the presence of a third molecule M (any air molecule) that removes the energy released by this reaction.

The destruction of ozone occurs by two different processes. One process $O_3 + h\nu_2 \rightarrow O_2 + O$ is the photodissociation under the influence of ultraviolet and visible light with energy larger than the binding energy $h\nu_2^e$ of O_3. The other process $O + O_3 \rightarrow O_2 + O_2$ is the destruction of ozone by collision with oxygen atoms.

The rates at which oxygen and ozone molecules are formed are thus given by

$$\frac{d[O_2]}{dt} = -k_1[O_2][h\nu_1 \geq h\nu_1^e]$$

$$\frac{d[O_2]}{dt} = -\frac{d[O_3]}{dt} = -k_2[O][O_2][M] \qquad (A.36)$$

$$\frac{d[O_2]}{dt} = -\frac{d[O_3]}{dt} = k_3[O_3][h\,\nu_2 \geq h\,\nu_2^e]$$

$$\frac{d[O_2]}{dt} = -2\frac{d[O_3]}{dt} = 2\,k_4[O][O_3]$$

or, when combined, by

$$\frac{d[O_2]}{dt} = -k_1[O_2][h\,\nu_1 \geq h\,\nu_1^e] - k_2[O][O_2][M]$$
$$+ k_3[O_3][h\,\nu_2 \geq h\,\nu_2^e] + 2\,k_4[O][O_3] \qquad \text{(A.37)}$$

$$\frac{d[O_3]}{dt} = k_2[O][O_2][M] - k_3[O_3][h\,\nu_2 \geq h\,\nu_2^e] - k_4[O][O_3]$$

These rates contribute to the source terms in the balance equations for oxygen and ozone. They depend not only on the densities $[O_2]$ and $[O_3]$ but also on the densities $[O]$ and $[M]$ and on the densities of photons with energies above certain thresholds.

The rates at which photons are absorbed are given by

$$\frac{d[h\,\nu_1 \geq h\,\nu_1^e]}{dt} = -k_1[O_2][h\,\nu_1 \geq h\,\nu_1^e]$$

$$\frac{d[h\,\nu_2 \geq h\,\nu_2^e]}{dt} = -k_3[O_3][h\,\nu_2 \geq h\,\nu_2^e] \qquad \text{(A.38)}$$

These rates contribute to the source term in the radiation balance equation. They do not only depend on the densities $[h\,\nu_1 \geq h\,\nu_1^e]$ and $[h\,\nu_2 \geq h\,\nu_2^e]$ but also on the densities $[O_2]$ and $[O_3]$.

This example shows again that the radiation balance equation depends on the concentrations of the radiatively active constituents of the atmosphere. In addition it shows that the balance equations for these active constituents depend on the "concentrations" of radiation in various frequency bands. The radiation and constituent balance equations are coupled.

This example also concludes our discussion of the basic fluid dynamical laws. Everything up to this point is well established and experimentally proven. It is basic physics. In the next three sections we consider the third closure problem and some of the customarily employed parameterizations of subgridscale fluxes. These parameterizations are based on a mixture of heuristic reasoning, dimensional analysis and formal analogues. These parameterizations are "educated guesses". We will not be on safe grounds any more.

A.6 Reynolds Decomposition

The balance equations, thermodynamic relations, and phenomenological flux laws describe fluid motions on all scales, from micro turbulence to wind systems, ocean circulations and ice sheet flows. These scales cannot be resolved simultaneously, neither observationally nor computationally. In tidal and climate modeling we only deal with large-scale or space-time averaged fields. Such averaged fields may be defined

$$\overline{\psi}(\boldsymbol{x},t) = \iiint d^3x' \int dt'\, \psi(\boldsymbol{x}',t')\, H(\boldsymbol{x}-\boldsymbol{x}',t-t') \qquad (A.39)$$

Formally this is a convolution between the field $\psi(\boldsymbol{x},t)$ and a filter function $H(\boldsymbol{x},t)$[5]. The averaging operation is linear and commutes with differentiation.

To obtain the equations for the averaged fields one decomposes all field variables $\psi(\boldsymbol{x},t)$ into a mean part $\overline{\psi}(\boldsymbol{x},t)$ and a fluctuating part $\psi'(\boldsymbol{x},t)$

$$\psi(\boldsymbol{x},t) = \overline{\psi}(\boldsymbol{x},t) + \psi'(\boldsymbol{x},t) \qquad (A.40)$$

This decomposition is called Reynolds decomposition[6]. It is assumed that the relations $\overline{\overline{\psi}} = \overline{\psi}$ and $\overline{\psi_1\psi_2} = \overline{\psi_1}\,\overline{\psi_2} + \overline{\psi_1'\psi_2'}$ hold for this Reynolds decomposition although this is not strictly true but an (generally good) approximation only[7].

How does the Reynolds decomposition (A.40) affect the dynamical equations? To answer this question consider first a simple quadratic equation

$$\partial_t\psi + a\psi + b\psi\psi = 0 \qquad (A.41)$$

Substituting the Reynolds decomposition, averaging and applying the rules of Reynolds averaging results in

$$\partial_t\overline{\psi} + a\overline{\psi} + b\overline{\psi}\,\overline{\psi} + b\overline{\psi'\psi'} = 0 \qquad (A.42)$$

Due to the quadratic nonlinearity the equation for $\overline{\psi}$ is not closed. It depends on the fluctuating part ψ' through the term $\overline{\psi'\psi'}$. Subtraction of equation (A.42) from (A.43) gives the equation for the fluctuating part

$$\partial_t\psi' + a\psi' + 2b\overline{\psi}\psi' + b\psi'\psi' - b\overline{\psi'\psi'} = 0 \qquad (A.43)$$

The equation for ψ' is also not closed. It depends on $\overline{\psi}$. The equations for $\overline{\psi}$ and ψ' are coupled through the nonlinear term. Nonlinearities couple different scales of motion. The Reynolds decomposition of equation (A.41) into

---

[5] Note that the averaging interval, *i.e.* the effective width of $H(\boldsymbol{x},t)$ determines which scales contribute to the mean part and which scales contribute to the fluctuating part. Scales that were part of the fluctuating field become part of the mean field when the averaging width is reduced.

[6] Such a decomposition is also performed in other fields, notably in statistical mechanics. There the decomposition refers to an ensemble of states Ψ, i.e., to a random process. Thus, averaging is not done across space and time, but across an ensemble of realizations, weighted by the frequency of the events. Then $\overline{\psi} = \mathrm{E}(\psi)$ and $\psi' = \psi - \mathrm{E}(\psi)$, with $\mathrm{E}(\cdot)$ being the expectation operator (cf. Appendix C.2.1).

[7] This assumption is based on the analogous decomposition of an ensemble, where $\mathrm{E}(\mathrm{E}(\psi)) = \mathrm{E}(\psi)$ and $\mathrm{Cov}(\psi_1,\psi_2) = \mathrm{E}(\psi_1)\mathrm{E}(\psi_2) + \mathrm{Cov}(\psi_1',\psi_2')$ hold strictly.

(A.42) and (A.43) does not cause any loss of information. The equation for the complete field ψ can be reestablished by adding (A.42) and (A.43).

If one is interested only in $\overline{\psi}$ then one does not need ψ' but $\overline{\psi'\psi'}$. Its equation is

$$\partial_t \overline{\psi'\psi'} + 2a\overline{\psi'\psi'} + 4b\overline{\psi}\,\overline{\psi'\psi'} + 2b\overline{\psi'\psi'\psi'} = 0 \tag{A.44}$$

and contains a triple product $\overline{\psi'\psi'\psi'}$. The time evolution of this triple product is governed by an equation that contains a fourth-order product, etc. One obtains an infinite hierarchy of coupled equations. To obtain a closed system one has to truncate this hierarchy by a closure hypothesis. Closure at order n assumes that all products of order $n+1$ and higher are expressed in terms of products of order m smaller or equal to n. A first order closure thus expresses $\overline{\psi'\psi'}$ in (A.42) in terms of $\overline{\psi}$.

The fluid dynamical equations contain nonlinearities in various places. Most important is the advective nonlinearity in the balance equations. Other nonlinearities arise because the thermodynamic coefficients and the molecular diffusion coefficients depend on (p, T, c_2). Here we only consider the effect of Reynolds decomposition and averaging on the advective nonlinearity. This effect depends on the choice of dependent variables. If we formulate the balance equations in terms of the total mass density ρ, the mass density ρ_2 of the second component, the momentum density $\boldsymbol{m} = \rho\boldsymbol{u}$, and the density ρe_t of the total energy then the advective flux has the form $\psi\boldsymbol{u}$ where ψ is either of the densities. The advective flux then decomposes into

$$\overline{\psi\boldsymbol{u}} = \overline{\psi}\,\overline{\boldsymbol{u}} + \overline{\psi'\boldsymbol{u}'} \tag{A.45}$$

where

$$\overline{\boldsymbol{u}} = \frac{\overline{\boldsymbol{m}} - \overline{\rho'\boldsymbol{u}'}}{\overline{\rho}} \tag{A.46}$$

In this representation the fluctuating components give rise to two effects. They cause an additional flux $\overline{\psi'\boldsymbol{u}'}$ in the balance equations, called the eddy or turbulent flux, and an additional contribution to the mean velocity

$$\overline{\boldsymbol{u}_e} = -\frac{\overline{\rho'\boldsymbol{u}'}}{\overline{\rho}} \tag{A.47}$$

called the eddy or turbulence induced mean velocity. The qualifier "eddy" or "turbulent" is used because the fluctuating fields are viewed as turbulent eddies. The eddy fluxes appear in the averaged balance equations for the same reason that molecular fluxes appeared in the original equations. There, a property was transported by a velocity that differed from the fluid velocity. The difference had to be accounted for by a molecular flux. Here, a property is transported by the velocity $\boldsymbol{u}$ which differs from the mean velocity $\overline{\boldsymbol{u}}$. The difference has to be accounted for by the eddy flux. Note that an eddy flux $\overline{\rho'\boldsymbol{u}'}$ appears in the continuity equation although it does not contain any molecular fluxes. The eddy-induced mean velocity $\boldsymbol{u}_e$ appears because the velocity is a nonlinear function of the state variables $(\rho, \rho c_2, \boldsymbol{m}, \rho e_t)$ given by $\boldsymbol{u} = \boldsymbol{m}/\rho$.

If one chooses to formulate the balance equations in terms of the state variables $(\rho, c_2, \boldsymbol{u}, e_t)$ then Reynolds decomposition of the material derivative leads to

$$\overline{\frac{D\chi}{Dt}} = \partial_t \overline{\chi} + \overline{\boldsymbol{u}} \cdot \boldsymbol{\nabla} \overline{\chi} + \overline{\boldsymbol{u}' \cdot \boldsymbol{\nabla} \chi'} \qquad (A.48)$$

where χ is any of the variables $(\rho, c_2, \boldsymbol{u}, e_t)$. Only the eddy terms $\overline{\boldsymbol{u}' \cdot \boldsymbol{\nabla} \chi'}$ appear. There is no eddy-induced mean velocity since $\boldsymbol{u}$ is one of the state variables. There arise, however, new nonlinearities on the right hand side of the equations. The pressure term, which is the linear expression $-\boldsymbol{\nabla} p$ in the equation for $\mathbf{m}$ becomes the nonlinear expression $-\rho^{-1} \boldsymbol{\nabla} p$ in the equation for $\boldsymbol{u}$.

In addition to the choice of state variables, the detailed effects of Reynolds averaging also depend on the choice of independent variables (height versus isopycnal, isobaric or isentropic coordinates), and on the form of the balance equations (flux form $\partial_t(\rho\chi) + \boldsymbol{\nabla} \cdot (\rho\chi\boldsymbol{u}) = \ldots$ versus advective form $\partial_t \chi + \boldsymbol{u} \cdot \boldsymbol{\nabla} \chi = \ldots$). In all cases there appear eddy-induced terms in the equations for the mean quantities. These eddy-induced terms represent the dynamic effect of the fluctuating fields on the mean fields. The specification of these eddy-induced terms represents the third closure problem.

The dynamical equations that form the basis of simulation models, such as Laplace tidal equations 4.2 or the equations 4.16 and 4.18 for the general circulation of the atmosphere and ocean, are such Reynolds-averaged equations. They are only valid for certain space and time scales. The "space-time" averaging (A.39) defines the scales that are explicitly modeled and the scales whose effect on the resolved scales must be parameterized. Next we discuss the most common parameterizations of eddy-induced fluxes: eddy diffusivities and eddy viscosities in the interior of the fluid and drag laws and Ekman suction at the boundary.

A.7 Parameterization of Interior Fluxes

Here we make the distinction between the eddy fluxes of scalar quantities such as temperature and salinity (specific humidity) and eddy fluxes of vector quantities such as momentum.

A.7.1 Eddy Diffusivities

In the interior of the ocean and atmosphere the eddy fluxes are often modeled after the molecular fluxes. One thus assumes

$$\overline{\psi' u_i'} = -K_{ij} \partial_j \overline{\psi} \qquad (A.49)$$

for the eddy fluxes of scalars ψ^8. Here $\boldsymbol{K}$ is the eddy (or turbulent) diffusion tensor. Equation (A.49) constitutes a parameterization of the eddy fluxes $\overline{\psi'u_i'}$.

Despite the general use of the parameterization (A.49) it is paramount to observe that there exists no general theory or framework that justifies its form and allows the calculation of $\boldsymbol{K}$. A heuristic estimate is often based on the mixing length theory which assumes

$$K \sim Lv' \tag{A.50}$$

where the mixing length L is a fluid analogue of the mean free path of molecules in a gas and v' a characteristic value of the turbulent velocity fluctuations. Furthermore, the diffusion tensor K_{ij} does not need to be constant. It can vary in space and time. It also does not need to be positive definite. There can be up-gradient fluxes. No general values can be assigned to the components of $\boldsymbol{K}$. The values will depend, among other things, on the scale that separates resolved from unresolved motions.

The values should, however, be the same for different tracers ψ. A tracer is a substance that moves with the fluid velocity, and the unresolved eddy currents should not distinguish between different tracers. Thus the eddy diffusion coefficients for potential temperature and salinity (or specific humidity) are assumed to be the same, in contrast to the molecular diffusion coefficient, which are different.

Usually one neglects molecular diffusion when eddy diffusion is introduced since molecular diffusion represents the transport by microscopic molecular motions, whereas eddy diffusion represents the larger transport by macroscopic turbulent motions.

The eddy diffusion tensor $\boldsymbol{K}$ can be decomposed into its symmetric component $\boldsymbol{K}^s(K_{ij}^s = K_{ji}^s)$ and antisymmetric component $\boldsymbol{K}^a(K_{ij}^a = -K_{ji}^a)$. When this decomposition is substituted into the divergence of the eddy flux one obtains

$$\partial_i(K_{ij}\partial_j\overline{\psi}) = v_j\partial_j\overline{\psi} + \partial_i(K_{ij}^s\partial_j\overline{\psi}) \tag{A.51}$$

where

$$v_j = \partial_i K_{ij}^a \tag{A.52}$$

The antisymmetric component thus acts like a mean velocity.

The orientation of the symmetric part $\boldsymbol{K}^s$ is an important issue. The principal axes may lie in or orthogonal to the geopotential, isopycnal, neutral,

---

[8] We use index notation in this section. The subscripts $i, j, \ldots$ run from 1 to 3 and denote the Cartesian components of a vector or tensor. Thus a_i $(i = 1, 2, 3)$ denotes the three Cartesian components of a three-dimensional vector $\mathbf{u}$ and T_{ij} $(i, j = 1, 2, 3)$ denotes the nine Cartesian components of a three-dimensional second-order tensor $\mathbf{T}$. An index that occurs once in an expression is called a *free index* and can take any value in its range. An index that occurs twice is called a *dummy index* and is summed over its values (summation convention, thus $a_i b_i = \sum_{i=1}^{3} a_i b_i$). An index cannot occur three times in an expression.

or any other surface. Most numerical models assume a diffusivity tensor that is diagonal in the horizontal/vertical coordinate system

$$\boldsymbol{K}^s = \begin{pmatrix} K_h & 0 & 0 \\ 0 & K_h & 0 \\ 0 & 0 & K_v \end{pmatrix} \tag{A.53}$$

with vertical diffusivity K_v (representing vertical mixing by small-scale eddies) much smaller than the horizontal coefficient K_h (representing horizontal stirring by meso-scale eddies). However stirring contributes to mixing across isopycnal surfaces

$$K_d = K_v + K_h s^2 \tag{A.54}$$

where K_d is the diapycnal (or cross-isopycnal) diffusion coefficient and s (assumed to be much smaller than 1) is the slope of the isopycnal. Since eddies stir primarily along isopycnal surfaces, some researchers argue that the diffusion tensor should be diagonal in an isopycnal/diapycnal coordinate system. Additional support for this representation seems to come from the argument that the exchange of particles on isopycnal surfaces does not require any work against gravity. However, neither of these arguments is fully convincing. Eddies can mix properties across *mean* isopycnal surfaces; and exchange of particles on horizontal, *i.e.*, geopotential surfaces, also does not require any work. Furthermore, potential energy is released when parcels are exchanged within the wedge between the horizontal and isopycnal surfaces. This led Olbers and Wenzel [126] to suggest a diffusion tensor with three principal components: a large value along the axis halfway between the isopycnal and geopotential surface where exchange would release the maximum amount of energy; a medium value along the intersection of the isopycnal and geopotential surfaces, where exchange results in no release of energy; and a small value along the axis perpendicular to these two axes, where mixing requires the maximum amount of work.

A.7.2 Eddy Viscosities

The eddy fluxes of momentum are parameterized by decomposing

$$\overline{\rho u_i' u_j'} = \rho \frac{\overline{u_k' u_k'}}{3} \delta_{ij} - \tau_{ij} \tag{A.55}$$

The first term on the right hand side can be absorbed into the pressure. The second term has zero trace and is parameterized by

$$\tau_{ij}/\rho = 2 A_{ijkl} \overline{D}_{kl} \tag{A.56}$$

where $\overline{D}_{kl}$ are the components of the mean rate of strain or deformation tensor (see Sect. A.1.3), and A_{ijkl} are the components of an eddy (or turbulent) viscosity tensor. $\mathbf{A}$ is a fourth order tensor and has 81 components. This number can be reduced by assuming that

(i) A_{ijkl} is symmetric in the first two and last two indices because of its definition (A.56).

(ii) $A_{iikl} = 0$ since $\tau_{ii} = 0$ because of (A.55)

(iii) $A_{ijkl} = A_{klij}$ because of "energy considerations", and

(iv) A_{ijkl} is axial-symmetric about the vertical axis, as we assumed in (A.53) for the diffusion tensor $\boldsymbol{K}$.

Under these assumptions the components of the stress tensor in (A.56) take the form

$$\tau_{11}/\rho = A_h(\overline{D}_{11} - \overline{D}_{22}) + A^* \left(\frac{\boldsymbol{\nabla} \cdot \boldsymbol{u}}{3} - \overline{D}_{33} \right)$$

$$\tau_{12}/\rho = \tau_{21}/\rho = 2A_h\overline{D}_{12}$$

$$\tau_{13}/\rho = \tau_{31}/\rho = 2A_v\overline{D}_{13} \qquad\qquad (A.57)$$

$$\tau_{22}/\rho = A_h(\overline{D}_{22} - \overline{D}_{11}) + A^* \left(\frac{\boldsymbol{\nabla} \cdot \boldsymbol{u}}{3} - \overline{D}_{33} \right)$$

$$\tau_{23}/\rho = \tau_{32}/\rho = 2A_v\overline{D}_{23}$$

$$\tau_{33}/\rho = -2A^* \left(\frac{\boldsymbol{\nabla} \cdot \boldsymbol{u}}{3} - \overline{D}_{33} \right)$$

with only three independent coefficients: the horizontal eddy viscosity coefficient A_h, the vertical eddy viscosity coefficient A_v, and the eddy viscosity coefficient A^*. If we had assumed isotropy we would have obtained $\tau_{ij} = 2A(D_{ij} - \frac{\boldsymbol{\nabla} \cdot \boldsymbol{u}}{3})$ with only one eddy viscosity coefficient. This differs from the result (A.21) for the molecular viscosity since there we did not require $\sigma_{ii} = 0$.

The actual values of the eddy viscosity coefficients depend on the scale that is chosen to separate resolved from unresolved motions. When this scale falls within the inertial range of three-dimensional homogeneous isotropic turbulence it can be shown that the effect of the unresolved on the resolved scales is given by an eddy viscosity coefficient proportional to the rate of strain tensor of the resolved scales [87]. Smagorinsky [153] applied analogous arguments to horizontally isotropic hydrostatic geophysical flows and derived

$$A^* = 0$$

$$A_h = c(\Delta x_h)^2 \left[(\overline{D}_{11} - \overline{D}_{22})^2 + (2\overline{D}_{12})^2 \right]^{1/2} \qquad (A.58)$$

$$A_V = c(\Delta x_v)^2 \left[(2\overline{D}_{13})^2 + (2\overline{D}_{23})^2 \right]^{1/2}$$

where Δx_h and Δx_v are the horizontal and vertical grid size and c is a constant. The fact that these eddy viscosity coefficients are not constant but depend on the rate of strain tensor of the resolved scales is referred to as nonlinear viscosity. The parameterization (A.58) is the most widely used parameterization in atmospheric general circulation models.

A.8 Parameterization of Boundary Layer Fluxes

As discussed in Sect. A.4, the boundary conditions for the actual fields at the actual interface are continuity conditions for the fluxes, velocities, temperatures and chemical potentials. When applying a Reynolds decomposition, these boundary conditions become highly nonlinear since the surface itself must also be decomposed into a mean and fluctuating part. Reynolds averaging then leads to many, generally intractable "eddy" terms. Instead of Reynolds averaging, coarse-resolution models employ boundary layer theory for the formulation of appropriate boundary conditions. Momentum, heat and water can be transported across a boundary only by molecular processes. They are first transported by molecular diffusion, within a thin molecular boundary layer, and then farther away by turbulent eddies. Within a constant flux layer the molecular and turbulent fluxes are constant and equal to the fluxes across the boundary. Farther away these quantities are deposited within a planetary boundary layer. This layer sets boundary conditions of fluid injection or removal for the interior of the fluid. In the atmosphere, the constant flux layer has a height of $O\,(10\,\mathrm{m})$ and the planetary boundary layer of $O\,(1\,\mathrm{km})$. The respective oceanic depths are $O\,(.05\,\mathrm{m})$ and $O\,(50\,\mathrm{m})$. In general, processes at and near boundaries are not yet sufficiently understood to provide unambiguous boundary conditions for the interior Reynolds-averaged flow in all circumstances. Some of the momentum and energy might escape into the interior in the form of internal waves. Topographic features that cannot be incorporated as bottom roughness complicate any "flat" bottom approach. Thus different schemes are used for different models, depending on resolution and other circumstances.

A.8.1 The Constant Flux Layer

Consider the atmosphere or ocean above a rigid horizontal surface and the eddy momentum flux components $\boldsymbol{\tau}_h = (\tau_{13}, \tau_{23})$ that describe the vertical, i.e., normal flux of horizontal momentum. Within the turbulent part of the constant flux layer, the velocity shear can only depend on the fluid density, the stress and, of course, the distance from the boundary. Dimensional arguments then demand the "law of the wall" with a logarithmic velocity profile and a drag law

$$\boldsymbol{\tau}_h = C_d \rho |\overline{\boldsymbol{u}}_h| \overline{\boldsymbol{u}}_h \qquad (A.59)$$

where $\overline{\boldsymbol{u}}_h = (\overline{u}_x, \overline{u}_y)$ is the mean horizontal velocity at some prescribed height and C_d a dimensionless drag coefficient, which depends on the bottom roughness. Above the air–sea interface the velocity in equation (A.59) has to be replaced by the velocity relative to the surface velocity $\overline{\boldsymbol{u}}_h(0)$, the surface drift current. The surface roughness and hence the drag coefficient then depend on the sea state, which in turn depends on the wind speed and, as the scatter in Fig. 2.2 (Sect. 2.5) indicates, on other not fully identified parameters.

Similarly, the flux of sensible heat across the air–sea interface and the evaporation rate are usually parameterized by

$$Q = \rho c_p c_H |\overline{\boldsymbol{u}}_h^a - \overline{\boldsymbol{u}}_h(0)| \left(\overline{T}^s - \overline{T}^a \right)$$
$$E = \rho c_E |\overline{\boldsymbol{u}}_h^a - \overline{\boldsymbol{u}}_h(0)| \left(\overline{q}^s - \overline{q}^a \right) \tag{A.60}$$

where $\overline{T}^a$ and $\overline{q}^a$ are the temperature and specific humidity at the same prescribed height within the atmospheric constant flux layer, $\overline{T}^s$ and $\overline{q}^s$ the temperature and specific humidity at the surface, and c_H and c_E dimensionless coefficients similar to the drag coefficient, often called the Stanton and the Dalton number.

If one assumes the logarithmic layer to be of infinitesimal thickness then the drag law (A.59) provides a boundary condition for the interior flow; the stress $\boldsymbol{\tau}_h$ is being specified by the velocity $\overline{\boldsymbol{u}}_h$ at the boundary. This assumption is usually made for atmospheric models that resolve the planetary boundary layer. The velocity used in the drag law is the one calculated at the lowest model level, nominally at zero height but assumed to represent the velocity at $O(2\,\mathrm{m})$ height. The drag law then provides the needed boundary condition for the atmosphere above land and the needed boundary conditions for the atmosphere and ocean at the air–sea interface.

A minor problem with the above scheme at the air–sea interface is that the drag law contains the surface drift current. This problem could be dealt with by assuming that the drag law (A.59) holds for both media and that the stress is the same. Then $C_d^a \rho^a |\boldsymbol{u}_h^a(z_*^a) - \boldsymbol{u}_h(0)|(\boldsymbol{u}_h^a(z_*^a) - \boldsymbol{u}_h(0)) = -C_d^o \rho^o |\boldsymbol{u}_h^o(z_*^o) - \boldsymbol{u}_h(0)|(\boldsymbol{u}_h^o(z_*^o) - \boldsymbol{u}_h(0))$ where the superscripts "a" and "o" denote the atmosphere and ocean, $\boldsymbol{u}_h(0)$ is again the mean surface drift velocity, and z_* is the height or depth of the constant flux layer. Continuity of the momentum flux $\boldsymbol{\tau}_h$ and of the drift velocity $\boldsymbol{u}_h(0)$ is assumed. This expression can be solved to give the mean drift velocity

$$\boldsymbol{u}_h(0) = \boldsymbol{u}_h^o(-z_*^o) + \frac{\epsilon}{1+\epsilon} \left(u_h^a(z_*^a) - u_h^o(-z_*^o) \right) \tag{A.61}$$

where $\epsilon = (C_d^a \rho^a / C_d^o \rho^o)^{1/2}$. Since the "kinematic" drag coefficients c_D^a and c_D^o are about equal one finds $\epsilon \approx (\rho^a/\rho^o)^{1/2} \approx 3 \times 10^{-2} \ll 1$. The drift velocity hence does not differ very much from the ocean velocity just outside the constant flux layer. The drift velocity (A.61) can be substituted into the drag law. The stress is then fully determined by velocities outside the boundary layers. Well-posed problems for both interior flows are obtained. The only problem with this approach is that the layer just below the ocean surface is not a simple log layer, but it is complicated by the effects of wave breaking. The usual way out is to simply assume the ocean or drift velocity to be zero, i.e., much smaller than the wind velocity.

A.8.2 The Planetary Boundary Layer

In both the atmosphere and ocean, the stress is distributed over a turbulent region of finite thickness, the planetary boundary layer. The typical height of the atmospheric planetary boundary layer is about 1km and usually well resolved in atmospheric models. The oceanic planetary boundary layer depth (below the surface) or height (above the bottom) is about $50\,\mathrm{m}$ and often not sufficiently resolved. Planetary boundary layer theory may then be used to convert the actual boundary conditions into boundary conditions for the inviscid interior. One prominent example is Ekman's [34] theory. Bulk mixed layer models are another example. We discuss both cases.

We present *Ekman's theory* first for the oceanic and atmospheric planetary boundary layer below and above the air–sea interface, although usually one only needs the results for the oceanic part. Then we consider the Ekman theory for the layer above the ocean bottom.

For the *atmosphere/ocean boundary* it is postulated that the divergence of the eddy stress is balanced in both boundary layers by the Coriolis force

$$\mathbf{f}_0 \times \boldsymbol{u}_h^{a,o} = \frac{1}{\rho_0^{o,a}} \frac{\partial}{\partial z} \boldsymbol{\tau}_h^{o,a} \tag{A.62}$$

where again the superscripts "o" and "a" denote the ocean and the atmosphere. The horizontal or Ekman transports in each fluid are then given by

$$
\begin{aligned}
\mathbf{M}^a &= \int_0^\infty \boldsymbol{u}_h^a &= \frac{1}{\rho_0^a f_0}(-\tau_y(0), \tau_x(0)) \\
\mathbf{M}^o &= \int_{-\infty}^0 \boldsymbol{u}_h^o &= \frac{1}{\rho_0^o f_0}(\tau_y(0), -\tau_x(0))
\end{aligned}
\tag{A.63}
$$

where use has been made that $\boldsymbol{\tau}_h$ is continuous at the interface $z = 0$ and vanishes for $z \to \pm\infty$. The Ekman transports are perpendicular to the surface stress and cancel each other, $\rho_0^a \mathbf{M}^a + \rho_0^o \mathbf{M}^o = 0$. Divergence or convergence of these Ekman transports causes vertical transport at the interface. Their magnitude is obtained from an integration of the "incompressibility condition" $\boldsymbol{\nabla}_h \cdot \boldsymbol{u}_h + \partial_z w = 0$ which yields

$$w^{a,o}(0) = \pm\boldsymbol{\nabla}_h \cdot \mathbf{M}^{a,o} \tag{A.64}$$

if again $w^{a,o}$ is assumed to vanish for $z \to \pm\infty$. These vertical transports must be balanced by the inviscid interior flow. The inviscid interior flow is thus subjected to an Ekman pumping or suction velocity

$$w_{Ek}^{a,o}(0) = -w^{a,o}(0) = \frac{1}{\rho_0^{a,o} f_0}(\partial_x \tau_y - \partial_y \tau_x) \tag{A.65}$$

that forces fluid into or out of the interior. The stress boundary condition for the complete flow is thus converted to a boundary condition for the normal velocity of the inviscid interior flow which injects or removes fluid from the interior. Oceanic models that do not resolve the Ekman layer often use this Ekman suction or pumping velocity as a boundary condition for the interior flow, with $\boldsymbol{\tau}_h$ given by the atmospheric drag law.

In the boundary layer at the *sea floor* the stress is not known. Ekman theory assumes that it can be parameterized as

$$\boldsymbol{\tau}_h(z) = \rho A_v \partial_z \boldsymbol{u}_h^b \tag{A.66}$$

where A_v is a constant vertical eddy viscosity and $\boldsymbol{u}_h^b$ the velocity in the boundary layer. The condition that the total flow be zero at the boundary then yields the surface stress

$$\tau_x(0) = \frac{\rho A_v}{\sqrt{2} h_{Ek}} (u_x - u_y) \quad \text{and} \quad \tau_y(0) = \frac{\rho A_v}{\sqrt{2} h_{Ek}} (u_x + u_y) \tag{A.67}$$

where $h_{Ek} = \sqrt{A_v/f_0}$ is the Ekman depth and $\boldsymbol{u} = (u_x, u_y)$ the interior velocity just outside the Ekman layer[9]. Substituting (A.67) into (A.65) yields the suction velocity

$$w_{Ek} = \frac{h_{Ek}}{\sqrt{2}} (\partial_x u_x + \partial_y u_y + \partial_x u_y - \partial_y u_x) \tag{A.68}$$

Note that the Ekman suction velocity depends in this case on the value of the vertical eddy viscosity coefficient.

The Ekman velocities exchange fluid and its properties (heat, salinity, potential vorticity, ...) with the ocean interior. To specify the associated property fluxes one needs to know the profiles of the properties and suction velocity within the boundary layer. Often these are calculated from *bulk mixed layer models*, which assume that temperature, salinity and velocity are constant down or up to a depth or height h and then change (sometimes discontinuously) to their interior values. For the oceanic surface boundary layer, this mixed layer depth h is often inferred from the turbulent kinetic energy budget, assuming that energy input by the wind, convection, shear instability, and perhaps other processes is balanced by mixing, dissipation, entrainment and perhaps other processes. For the oceanic bottom boundary layer, the turbulent kinetic energy budget is more elusive and the mixed layer height is usually inferred from less founded turbulent scaling laws. The situation at the bottom is further complicated by the fact that the bottom is not flat but contains slopes and roughness elements on all scales.

Overall, atmospheric models usually use the quadratic drag law (A.59) to determine the stress at their lower boundary, with an appropriate drag

---

[9] The "linear" drag law (A.67) expresses the surface stress in terms of the velocity just outside the planetary boundary layer whereas the "quadratic" drag law (A.59) expresses the surface stress in terms of the velocity just outside the constant flux layer. Linear and quadratic bottom friction differ by more than a different coefficient in the drag law.

coefficient and wind velocity. At the surface, oceanic models apply the same stress, either directly or to determine the Ekman suction velocity or to drive a surface mixed layer model. At the bottom, oceanic models use a wide variety of boundary conditions, including the no-slip condition $\boldsymbol{u}_h = 0$, the quadratic drag law, the free-slip condition $\partial_z \boldsymbol{u}_h = 0$, the linear drag law with its associated Ekman suction velocity, or an explicit boundary layer based on rationales from turbulent theories.

It is again stressed that these prescriptions of the boundary conditions are parameterizations and should be treated with the same scepticism as the eddy diffusion and eddy viscosity coefficients.

A.9 Approximations

The dynamics resolved by the Reynolds averaged equations may contain processes that do not exert a strong influence on the problem to be studied. Such minor or irrelevant processes ought to be eliminated for transparency and, more importantly, for numerical efficiency. This elimination leads to approximations of the governing dynamical equations. Approximations differ from parameterizations in that one knows, in principle, how to reinstate the eliminated process. There are various types of approximations. We illustrate two of them that are relevant for large-scale atmospheric and oceanic flows.

A.9.1 Anelastic Approximation

If one disturbs a fluid flow at some point it responds or adjusts to the disturbance by emitting waves. The waves communicate the disturbance to other parts of the fluid. The most common types of waves are sound waves, gravity waves and Rossby waves. These waves have different restoring mechanisms: compressibility for sound waves, gravitation and stratification for gravity waves, and the rotation and sphericity of the earth for Rossby waves. The dispersion relations for these waves are shown in Fig. A.4. Since waves have a finite wave speed it takes time for the fluid to adjust to a disturbance. However, if this time is short compared to the time scales of interest one can assume that the adjustment is instantaneous. An example of this type of approximation is the anelastic approximation. Consider equation (A.15) for the pressure

$$\frac{D}{Dt}p = -\rho c^2 \, \boldsymbol{\nabla} \cdot \boldsymbol{u} - D \tag{A.69}$$

where D summarizes the dissipative terms. Decompose the pressure into a hydrostatically balanced background part $\tilde{p}(z)$ and a dynamically active part $p'(\boldsymbol{x}, t)$. The pressure equation then takes the form

$$\frac{D}{Dt}p' + w\frac{d}{dz}\tilde{p} = -\rho c^2 \boldsymbol{\nabla} \cdot \boldsymbol{u} - D \tag{A.70}$$

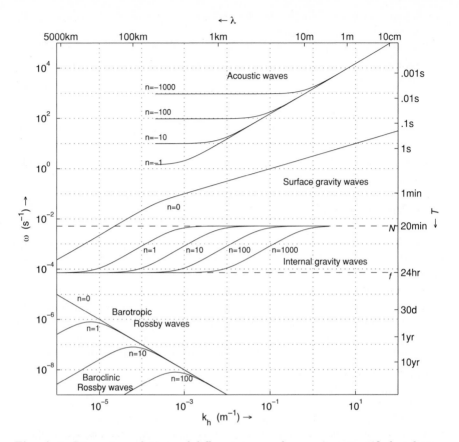

Fig. A.4. Dispersion relations of different types of waves in a stratified and compressible fluid layer. The *horizontal axis* represents horizontal wavenumbers (*bottom*) or horizontal wavelengths (*top*); the *vertical axis* represents frequency (*left*) or period (*right*). The different types of wave motions are acoustic waves (related to compressibility), surface and internal gravity waves (related to gravitation and stratification), and Rossby waves (related to earth rotation and sphericity). The "mode" number n characterizes the vertical structure of the waves. It is number of nodes within the fluid layer

Pressure disturbances are transmitted by sound waves at a speed of about $c \approx 1410\,\text{m s}^{-1}$ in water and $c \approx 330\,\text{m s}^{-1}$ in air. If one is interested in slower processes one can neglect the time derivative in (A.70) and obtains $w d\tilde{p}/dz = -\rho c^2 \, \boldsymbol{\nabla} \cdot \boldsymbol{u} - D$. This is a diagnostic equation for the velocity divergence $\boldsymbol{\nabla} \cdot \boldsymbol{u}$. In the ocean one can additionally neglect the advective term $w d\tilde{p}/dz$ and the dissipative term D with respect to any of the individual terms of the velocity divergence. The three terms of the velocity divergence must thus balance and the velocity field becomes non-divergent or solenoidal

$$\nabla \cdot \boldsymbol{u} = 0 \tag{A.71}$$

The dynamic part p' of the pressure is now determined by the divergence of the momentum balance

$$\nabla \cdot \left(\frac{1}{\rho} \nabla p' \right) = \tag{A.72}$$

$$-\nabla \cdot (\boldsymbol{u} \cdot \nabla \boldsymbol{u}) + 2\boldsymbol{\Omega} \cdot (\nabla \times \boldsymbol{u}) - \nabla \cdot \left(\frac{\rho'}{\rho} \nabla \phi \right) + \nabla \cdot \left(\frac{1}{\rho} \tau \right)$$

which is a three-dimensional Poisson equation for p'. Here ρ' is the dynamically active part of the density field. The essential effect of the neglect of the time derivative in (A.70) is the elimination of sound waves from the system. As a result the velocity divergence and the pressure adjust instantaneously and are given by the diagnostic equations (A.71) and (A.72), in the case of the ocean. Note that the fluid is still compressible. The equation of state still depends on the pressure. In the ocean one can further replace the density in the inertial terms by a constant reference density and then arrives at the Boussinesq approximation.

In the next section we show that the pressure for large-scale atmospheric and oceanic motions is actually determined by a much simpler equation than (A.72), namely the hydrostatic balance.

A.9.2 Shallow Water Approximation

Another approximation arises from the fact that the atmosphere and the ocean can be regarded as shallow fluid layers if the horizontal length scale L of the fluid motion is much larger than its vertical scale H. For such motions the vertical velocity is much smaller than the horizontal velocity and the vertical momentum balance can be approximated by the hydrostatic balance

$$\frac{\partial}{\partial z} p' = -\rho' g \tag{A.73}$$

where $g = |\nabla \phi|$ is the gravitational acceleration. Formally, the hydrostatic balance represents the lowest order of an asymptotic expansion of the vertical momentum balance with respect to the aspect ratio H/L. For consistency the hydrostatic approximation needs to be accompanied by other approximations.

Consider first the Coriolis vector. At each latitude φ it can be written $2\boldsymbol{\Omega} = (0, \tilde{f}, f)$ where $\tilde{f} = 2\Omega \cos \varphi$ is the local meridional component and $f = 2\Omega \sin \varphi$ the local vertical component. Both components appear in the horizontal momentum balances. Consistency with the hydrostatic balance, however, requires neglecting the meridional component. This is the traditional approximation. Among the two directions in the original problem, the gradient of the geopotential which defines the local vertical direction and the axis of the earth rotation aligned with $2\boldsymbol{\Omega}$, the local vertical direction turns out to be more relevant. Gravitation wins over rotation.

Also for consistency one needs to approximate the expression (A.57) for the eddy-induced stress tensor. First, one needs to set the eddy viscosity coefficient A^* to zero. Second, one needs to approximate

$$2\overline{D}_{13} = \partial u/\partial z$$
$$2\overline{D}_{23} = \partial v/\partial z \qquad (A.74)$$

The horizontal momentum balance thus reduces to

$$\frac{D}{Dt}\overline{\boldsymbol{u}}_h + f\hat{\boldsymbol{z}} \times \overline{\boldsymbol{u}}_h = -\frac{1}{\rho}\boldsymbol{\nabla}_h p + A_h \Delta_h \overline{\boldsymbol{u}}_h + \partial_z(A_v \partial_z \overline{\boldsymbol{u}}_h) \qquad (A.75)$$

with vertical and horizontal eddy viscosity coefficients A_v and A_h. On the spherical earth these and the other equations are usually written in spherical coordinates $(\theta, \varphi, z = r - r_0)$ where θ is longitude, φ latitude and z the vertical coordinate with reference to some constant radius r_0. A third approximation then assumes the metric of these coordinates to be given by

$$ds^2 = r_0^2 \cos^2 \varphi d\theta^2 + r_0^2 d\varphi^2 + dz^2 \qquad (A.76)$$

with the constant radius r_0 instead of the actual radius $r = r_0 + z$ in the coefficients.

These and some further approximation constitute the shallow water approximation and result in the shallow water equations. These consist of prognostic equations for the two horizontal velocity components u and v, the temperature T, and the concentration c_2. The density is given by the equation of state, the pressure by the hydrostatic approximation, and the vertical velocity by the constraint on the velocity divergence. They are diagnostic variables. General circulation models of the atmosphere and ocean are generally based on the shallow water equations, as discussed in Chap. 4. The shallow water approximation renders the problem more two-dimensional. This fact is utilized when the fluid is assumed to be a set of horizontal layers.

These and other approximations are systematically derived in textbooks of oceanography and meteorology (e.g., [129], [63], [45]).

A.10 Representations

Dynamical equations can be expressed in different coordinate systems and by different sets of dependent variables. One usually strives for coordinate systems and variables, *i.e.*, for representations, that are optimal in some sense. Processes on the spherical earth are best described in spherical coordinates. Other coordinate systems are not wrong but may be cumbersome. Less obvious are the advantages of other representations. We discuss two examples.

A.10.1 Vertical Coordinates

Instead of the vertical coordinate z one can introduce any other coordinate $\eta(x, y, z, t)$ as long as it is a monotonic and invertible function of z. The vertical coordinate z or the geopotential height $\phi = gz$ then becomes a dependent variable. Typical choices of η are the (potential) density, pressure, or specific entropy leading to the isopycnal, isobaric or isentropic coordinate systems. The transformation rules are well known. They imply the following:

The advective derivative for any scalar function g becomes $Dg/Dt = (\partial_t + u_h \cdot \nabla_h + \omega \partial/\partial \eta)g$, where $\omega = D\eta/Dt$ is the fluid velocity through the $\eta = const.$ surfaces. It replaces the vertical velocity. If η is a materially conserved tracer ($D\eta/Dt = 0$) then $\omega = 0$ and the advective operator becomes two-dimensional, a great simplification. In oceanography one often uses the isopycnal coordinate system with $\rho = \rho_{pot}$. In this case ω is only due to dissipative processes, again a substantial simplification.

The continuity equation can be written as $\frac{D}{Dt} ln\rho = -\nabla \cdot u$. When transformed it becomes $\frac{D}{Dt} \ln \rho \frac{\partial z}{\partial \eta} = -\frac{D}{Dt} \ln \frac{\partial z}{\partial \eta} - \nabla_h \cdot u_h - \frac{\partial \omega}{\partial \eta}$. If the hydrostatic balance $\partial p / \partial \eta = -\rho g \partial z / \partial \eta$ is used then

$$\frac{D}{Dt} \ln \frac{\partial p}{\partial \eta} = -\nabla_h \cdot u_h - \frac{\partial \omega}{\partial \eta} \tag{A.77}$$

If one chooses $\eta = p$ then the left hand side vanishes and the continuity equation becomes linear.

Similarly, the horizontal acceleration due to pressure forces becomes

$$\frac{1}{\rho} \nabla_h p|_z = \frac{1}{\rho} \nabla_h p|_\eta - \frac{1}{\rho} \frac{\partial p}{\partial \eta} \frac{\partial z}{\partial \eta} - \nabla_h z = \frac{1}{\rho} \nabla_h p + g \nabla_h z \tag{A.78}$$

For $\eta = p$ the first term on the right hand side vanishes and the pressure force becomes linear. Such linearizations, obtained simply by going from height coordinates to isobaric coordinates may simplify analyses and computations considerably. In oceanography, the continuity equation becomes linear in height coordinates when the anelastic approximation is applied (see (A.71)). The acceleration due to the pressure force becomes linear in height coordinates when one applies the Boussinesq approximation which allows replacing ρ by $\rho = const.$ in the inertial terms of the momentum balance and hence on the left hand side of (A.78).

A.10.2 Decoupling

Introducing new sets of dependent variables can also have its advantages. We demonstrate this by considering the linearized version of Laplace tidal equations

$$\partial_t u - f_0 v = -g \partial_x \xi - \partial_x \phi_T$$
$$\partial_t v + f_0 u = -g \partial_y \xi - \partial_y \phi_T \tag{A.79}$$
$$\partial_t \xi + h_0 (\partial_x u + \partial_y v) = 0$$

on the f-plane with no friction. Here h_0 is the constant depth, f_0 the constant Coriolis parameter and ϕ_T the tidal potential. The three dependent variables are the zonal velocity u, the meridional velocity v, and the surface elevation ξ. If one introduces the vorticity and divergence

$$\zeta = \partial_x v - \partial_y u$$
$$d = \partial_x u + \partial_y v \tag{A.80}$$

the equations reduce to

$$\partial_t \zeta + f_0 d = 0$$
$$\partial_t d - f_0 \zeta = -g \nabla_h^2 \xi - \nabla_h^2 \phi_T \tag{A.81}$$
$$\partial_t \xi + h_0 d = 0$$

If one further introduces the potential vorticity $q = \zeta - \frac{f_0}{h_0} \xi$ one obtains

$$\partial_t q = 0$$
$$\partial_t \partial_t d + (f_0^2 - g h_0 \nabla_h^2) d = -\nabla_h^2 \partial_t \phi_T \tag{A.82}$$

In this representation the equations for q and d are decoupled. The first equation describes steady currents which carry potential vorticity q. These currents are geostrophically balanced and horizontally non-divergent. They are not forced by the tidal potential. The second equation describes gravity waves which have zero potential vorticity and dispersion relation $\omega^2 = f_0^2 - g h_0 k^2$. They are forced by the tidal potential. The decoupling has the advantage that it separates these two different types of dynamics that are intermingled in the original equations. The solutions for each type can be constructed independently. The complete solution is then obtained by superposition.

This strict decoupling becomes invalid once nonlinearities, friction, variable water depth and the earth's sphericity are introduced. Nevertheless, even under these circumstances it might be advantageous to work with vorticity ζ and divergence d, instead of the velocity components u and v, since these variables are still governed by distinctively different physics. While tidal studies generally do not make use of the inherent advantages of such a representation much of the theory of the large scale oceanic and atmospheric circulation does. There, balance equations are derived for flows that are in gradient-wind balance and weakly divergent and whose dynamical evolution is solely governed by the potential vorticity equation. A representation that decouples the linear problem is also advantageous when a solution to the nonlinear problem is sought by a perturbation expansion about the solution of the linear problem.

B

Numerics

This appendix describes basic properties of the algorithms that are used to obtain numerical solutions of dynamical equations. The finite difference approximation and its truncation error are discussed first, in Sect. B.1. The most important property of any numerical solution of a differential equation is its stability. The stability of numerical solutions of partial differential equations depends on whether the equations are elliptic, parabolic or hyperbolic. These three cases are discussed separately in Sect. B.2. Equations describing environmental systems also contain more than one dependent variable. Different dependent variables can then be placed on different spatial grids. The most common of these staggered grids are described in Sect. B.3. Instead of spatial grids numerical models also employ spectral or finite element methods to represent the spatial structure. These are discussed in Sects. B.4 and B.5. More details about numerical methods in ocean and atmosphere modeling can be found in [1], [2], [51], [54], [76], [113], and [186].

B.1 Discretization

A digital computer can only manipulate a finite number of discrete pieces of information. Therefore a continuous function can only be reproduced at preselected discrete points. Derivatives must be replaced by finite differences. This introduces truncation errors.

Consider the function $u(t)$. The preselected points t_n, $n = 0, 1 \ldots N$, form a grid. Often an equi-spaced grid is used such that $t_n = n\Delta t$. Then Δt is called the grid spacing or the time step, if t denotes the time. The value of the function at the grid point is denoted by u_n. If this function satisfies the differential equation

$$\frac{du}{dt} = F(u) \tag{B.1}$$

then the derivative is obtained from the Taylor expansion

$$u_{n+1} = u_n + \left.\frac{du}{dt}\right|_n \Delta t + \frac{1}{2!}\left.\frac{d^2u}{dt^2}\right|_n (\Delta t)^2 + \dots \tag{B.2}$$

as

$$\left.\frac{du}{dt}\right|_n = \frac{u_{n+1} - u_n}{\Delta t} - \frac{1}{2!}\left.\frac{d^2u}{dt^2}\right|_n \Delta t + \dots \tag{B.3}$$

The first term is the finite difference approximation for the first derivative. The second and all higher terms are the truncation error. For this specific scheme, the truncation is $O(\Delta t)$. The scheme is said to be of first order accuracy. If one uses the Taylor expansion

$$u_{n-1} = u_n - \left.\frac{du}{dt}\right|_n \Delta t + \frac{1}{2!}\left.\frac{d^2u}{dt^2}\right|_n (\Delta t)^2 + \dots \tag{B.4}$$

then one obtains by subtraction

$$\left.\frac{du}{dt}\right|_n = \frac{u_{n+1} - u_{n-1}}{2\Delta t} + O((\Delta t)^2) \tag{B.5}$$

a finite difference approximation that is of second order accuracy. The finite difference approximation (B.3) is called forward (or downwind) differencing. The scheme (B.5) is called a centered difference. There is also a backward (or upwind) finite difference

$$\left.\frac{du}{dt}\right|_n = \frac{u_n - u_{n-1}}{\Delta t} + O(\Delta t) \tag{B.6}$$

which is of first order accuracy. Taylor expansion also provides finite difference approximations for higher order derivatives. Adding (B.2) and (B.4) gives

$$\left.\frac{d^2u}{dt^2}\right|_n = \frac{u_{n+1} - 2u_n + u_{n-1}}{(\Delta t)^2} + O((\Delta t)^2) \tag{B.7}$$

The truncation error is under control of the modeler. It can be reduced by decreasing the grid spacing Δt. Of course this will decrease the efficiency as a larger array of points has to be handled. Also, the larger number of operations increases the round-off errors introduced by the finite machine accuracy. Eventually, the round-off errors tend to become larger than the truncation errors and any further reduction of the grid spacing defeats its purpose.

The truncation error can also be reduced by increasing the order of the finite difference approximation. The trade-off here is that higher order schemes increase the complexity of the code especially when formulating the boundary conditions. For this reason, atmosphere and ocean models generally use second order or, at most, fourth order schemes.

When the finite difference approximations are substituted into the differential equation, one obtains the finite difference equation

$$\frac{du}{dt}\bigg|_n = F(u_n) \tag{B.8}$$

If the forward scheme (B.3) is implemented, we get the explicit equation

$$u_{n+1} = u_n + \Delta t F(u_n) \tag{B.9}$$

It explicitly gives the $(n+1)$th value in terms of the nth value. One can march through the solution from an initial value u_0. If one uses the backward scheme, then one obtains the implicit equation

$$u_{n+1} = u_n + \Delta t F(u_{n+1}) \tag{B.10}$$

To obtain u_{n+1} from u_n, one first has to invert F. This can be a major computational effort in geophysical problem, where F is generally a high dimensional operator. One can lessen the effort by using the semi-implicit equation

$$u_{n+1} = u_n + \Delta t \left[F(u_n) + \frac{dF}{du}\bigg|_n (u_{n+1} - u_n) \right] \tag{B.11}$$

where $F(u_{n+1})$ has been approximated by a Taylor expansion.

Finite difference approximations should also be consistent and convergent. A finite difference approximation is consistent with the original differential equation if the difference equation converges to the differential equation as the grid spacing Δt tends to zero. For finite difference approximations that are based on Taylor expansions this property is easily checked and satisfied. Consistency generally does not constitute a problem. The solution of a finite difference approximation is said to converge to the solution of the original continuous differential equation if the difference between the two solutions at a fixed point t tends to zero uniformly as $\Delta t \to 0$ and $n \to \infty$. Convergence is much harder to prove[1]. Generally one relies on the Lax Equivalence theorem. It states that consistency and stability imply convergence for linear systems.

The most important property of a finite difference approximation is its stability. Let u_n be the exact solution of the differential equation and U_n the solution of the finite difference approximation. The finite difference solution

---

[1] Note, however, that consistency and convergence of difference schemes allow only for an assessment of the performance of the difference scheme. These properties cannot be transferred to the underlying "physical" differential equations. As outlined in Sect. 2.1.3 these equations are only valid for phenomena on a certain scale. The character of these equations changes qualitatively when phenomena on other scales are considered. Subgridscale parameterizations depend on scale and change in a discontinuous manner with scale. The number of dependent variables may change, etc. If a difference equation converges to a differential equation as Δt tends to zero this equation is not the physical equation valid for arbitrarily small scales. Such an equation does not exist because, among other reasons, one leaves the realm of continuum physics and enters the realm of discrete particle physics.

is stable if the difference $U_n - u_n$ stays bounded as n tends to infinity for fixed Δt.

Consider the linear differential equation $\frac{du}{dt} = -\gamma u$, with initial value $u(t = 0) = u_0$ and $\gamma > 0$. It has the solution $u(t) = u_0 e^{-\gamma t}$, which tends to zero for $t \to \infty$. Using the forward difference or explicit scheme, we get $\frac{u_{n+1} - u_n}{\Delta t} = -\gamma u_n$, or $u_{n+1} = (1 - \gamma \Delta t) u_n$. For $\Delta t > 2/\gamma$ the value $|u_n|$ increases monotonically with n and the solution is unstable. For the backward or implicit scheme, we find $u_{n+1} = u_n - \Delta t \gamma u_{n+1}$, or $u_{n+1} = \frac{u_n}{1 + \gamma \Delta t}$. In this case $|u_n|$ decreases monotonically with n, independent of the grid size Δt. The implicit scheme is unconditionally stable. The semi-implicit method results in $u_{n+1} - \frac{1 - \gamma \Delta t - \gamma^2 \Delta t}{1 - \gamma^2 \Delta t} u_n$, which is again unconditionally stable.

B.2 Partial Differential Equations

Geophysical flows are governed by a set of partial differential equations. These PDEs contain nonlinear advection terms $\boldsymbol{u} \cdot \nabla \varphi$ which describe the advection of a fluid property φ by the fluid velocity $\boldsymbol{u}$. These nonlinear advection terms are the major cause that renders analytic solutions impossible. The highest order derivatives are generally diffusive terms of the form $D \nabla \cdot \nabla \varphi$. These highest order derivatives determine the general character of the PDE. This is usually demonstrated by the simple second order PDE

$$A \partial_t \partial_t u + B \partial_t \partial_x u + C \partial_x \partial_x u + F(\partial_t u, \partial_x u, u; x, t) = 0 \qquad \text{(B.12)}$$

for the scalar $u(x, t)$. The coefficients A, B and C of the highest order derivatives determine the characteristic directions in (x, t)-space $c_{\pm} = \frac{B \pm \sqrt{B^2 - 4AC}}{2A}$. These characteristics describe the lines in (x, t)-space along which information propagates. This propagation differs depending on the nature of these characteristics. We call the PDE

$$\begin{array}{lll} \text{hyperbolic if} & B^2 > 4AC \\ \text{parabolic} \quad \text{if} & B^2 = 4AC \\ \text{elliptic} \quad \text{if} & B^2 < 4AC \end{array}$$

Hyperbolic equations have two distinct real characteristics. Wave propagation $\partial_t \partial_t u - c^2 \partial_x \partial_x u = 0$ is a typical hyperbolic problem.

Parabolic equations have two coinciding real characteristics. The diffusion equation $\partial_t u = D \partial_x \partial_x u$ is the prototype example.

Elliptic equations have two imaginary characteristics. They usually occur for steady state problems. Laplace and Poisson equations are typical examples.

The distinction between elliptic, parabolic and hyperbolic systems also carries over to the PDEs that describe geophysical flows. These PDEs can be converted to a set of first order PDEs for a vector $\boldsymbol{y} = (y_1, \ldots, y_N)$ which is a function of position $\boldsymbol{x} = (x, y, z)$ and time t

$$\partial_t y_i + A_{ij}\partial_x y_j + B_{ij}\partial_y y_j + C_{ij}\partial_z y_j + F_i(y; x, t) = 0 \qquad (B.13)$$

The eigenvalues of the matrices $\boldsymbol{A}$, $\boldsymbol{B}$ and $\boldsymbol{C}$ then determine the character of the PDE. If the eigenvalues of $\boldsymbol{A}$ are real and distinct, the problem is hyperbolic in (x, t) space. If they are imaginary, the problem is elliptic in (x, t) space. Similarly, the eigenvalues of the matrices $\boldsymbol{B}$ and $\boldsymbol{C}$ determine the nature of the problem in (y, t) and (z, t) space.

Since the numerical algorithms depend on the nature of the PDE, we discuss elliptic, parabolic and hyperbolic equations separately.

B.2.1 Elliptic Problems

Elliptic equations arise in oceanography and meteorology for steady state flows, such as the steady wind-driven oceanic circulations studied by Stommel and Munk. They can also arise as part of a time-dependent problem. One example are oceanic flows under the Boussinesq but nonhydrostatic approximation. The Boussinesq approximation assumes the velocity field to be nondivergent or solenoidal, $\nabla \cdot \boldsymbol{u} = 0$. Mass conservation is replaced by volume conservation. To assure this condition in the nonhydrostatic approximation, one takes the divergence of the momentum equation $\partial_t \boldsymbol{u} + \boldsymbol{u} \cdot \nabla \cdot \boldsymbol{u} = -\frac{1}{\rho}\nabla p + \boldsymbol{F}$, and arrives at a Poisson equation for the pressure p

$$\nabla \cdot \left(\frac{1}{\rho}\nabla\right) p = \nabla \cdot (\boldsymbol{u} \cdot \nabla \boldsymbol{u}) + \nabla \cdot \boldsymbol{F} \qquad (B.14)$$

This equation has to be solved at any time step for given $\boldsymbol{u}$, $\boldsymbol{F}$ and ρ. An elliptic equation also arises for two-dimensional incompressible flows, $\partial_x u + \partial_y v = 0$. In this case a stream function ψ can be introduced, such that $u = -\partial_y\psi$ and $v = \partial_x\psi$, and is governed by the equation $\Delta\psi = \partial_x v - \partial_y u$, with the vertical component of the vorticity vector as the source term.

A well-posed elliptic problem requires specification of boundary conditions. These may be Dirichlet, von Neumann, mixed, or periodic. The boundary conditions affect the solution in the whole domain at once. Physically, any perturbation is felt immediately everywhere. The speed of adjustment is infinite. In geophysical flows, pressure perturbations are transmitted by sound waves. The Boussinesq approximation eliminates sound waves by assuming their speed to be infinite. As a consequence the pressure is determined instantaneously by the Poisson equation (B.14).

Most of the elliptic equations encountered in oceanography and meteorology are well posed linear problems for which it is easy to show that a unique solution exists, sometimes to within an additive constant. The numerical algorithms are designed to find these solutions accurately and efficiently.

For illustration consider the two dimensional Poisson equation

$$(\partial_x\partial_x + \partial_y\partial_y)u(x, y) = f(x, y) \qquad (B.15)$$

The centered difference approximation (B.7) results in $u_{i+1,j} - 4u_{i,j} + u_{i-1,j} + u_{i,j+1} + u_{i,j-1} = 2f_{i,j}$, where $i = 1, \ldots, I$ and $j = 1, \ldots, J$ and where $\Delta x = \Delta y = 1$ for simplicity. This set of linear equations can be written as the matrix equation

$$\boldsymbol{A} \cdot \boldsymbol{x} = \boldsymbol{b} \tag{B.16}$$

where $\boldsymbol{x}$ is a vector of length $N = I \cdot J$ which contains all the elements $u_{i,j}$, $\boldsymbol{b}$ is a vector of the same length which contains all the elements $f_{i,j}$ and $\boldsymbol{A}$ is the $N \times N$ coefficient matrix. The coefficient matrix is sparse (most elements are zero) and banded (essentially tridiagonal). To find the solution $u_{i,j}$ one thus has to invert the matrix $\boldsymbol{A}$. There are direct and iterative methods to do so.

Direct methods are all variants of the Gauss forward elimination procedure where all subdiagonal elements are eliminated by normalization and subtraction. The resulting upper triangular matrix is then solved by back-substitution. These variants go by the name of Gauss–Jordan elimination, LU decomposition and Thomas algorithm. For large dense matrices these algorithms are computationally inefficient. For sparse and banded matrices they can be made more efficient. However, for most geophysical problems, iterative or relaxation methods are used.

Iterative schemes are based on the fixed point theorem. For vector functions it states that the scheme

$$\boldsymbol{x}^{(m+1)} = \boldsymbol{M}\boldsymbol{x}^{(m)} + \boldsymbol{b} \tag{B.17}$$

converges to the fixed point $\boldsymbol{x} = \boldsymbol{M}\boldsymbol{x} + \boldsymbol{b}$ if the largest eigenvalue (i.e., the "spectral radius") of the matrix $\boldsymbol{M}$ is smaller than one. This can be seen by considering the error $\Delta \boldsymbol{x}^{(m)} = \boldsymbol{x} - \boldsymbol{x}^{(m)}$ which is governed by $\Delta \boldsymbol{x}^{(m+1)} = \boldsymbol{M}\boldsymbol{x}^{(m)} = \boldsymbol{M}^m \boldsymbol{x}^{(0)}$. The error thus converges to zero if $\boldsymbol{M}^m$ converges to zero which requires the eigenvalues of $\boldsymbol{M}$ or its spectral radius to be smaller than one. The convergence is the faster the smaller the eigenvalues. If $\boldsymbol{M} = \boldsymbol{I} - \boldsymbol{A}$ where $\boldsymbol{I}$ is the identity matrix, the fixed point solves the equation $\boldsymbol{A}\boldsymbol{x} = \boldsymbol{b}$.

To apply iterative schemes, one decomposes the matrix

$$\boldsymbol{A} = \boldsymbol{D} - \boldsymbol{L} - \boldsymbol{U} \tag{B.18}$$

where $\boldsymbol{D}$ is diagonal (or easily invertible) and $\boldsymbol{L}$ and $\boldsymbol{U}$ are lower and upper triangular matrices. The matrix equation (B.16) then takes the form $\boldsymbol{D}\boldsymbol{x} = (\boldsymbol{L} + \boldsymbol{U})\boldsymbol{x} + \boldsymbol{b}$. The *Gauss–Jacobi* iteration is given by

$$\boldsymbol{x}^{(m+1)} = \boldsymbol{D}^{-1}(\boldsymbol{L} + \boldsymbol{U})\boldsymbol{x}^{(m)} + \boldsymbol{D}^{-1}\boldsymbol{b} \tag{B.19}$$

the *Gauss–Seidel* iteration by

$$(\boldsymbol{D} - \boldsymbol{L})\boldsymbol{x}^{(m+1)} = \boldsymbol{U}\boldsymbol{x}^{(m)} + \boldsymbol{b} \tag{B.20}$$

and the *relaxation* scheme by

$$x^{(m+1)} = x^{(m)} + \alpha(\hat{x}^{(m+1)} - x^{(m)}) \tag{B.21}$$

where $\hat{x}^{(m+1)}$ on the right hand side is given by the expression (B.19). The parameter α is called the relaxation parameter. Convergence requires $0 < \alpha < 2$.

The major issue for these and other schemes is the rate of convergence. It can be shown that Gauss–Seidel is twice as fast as the Gauss–Jacobi scheme, and that the relaxation scheme is faster than the Gauss–Seidel scheme if $\alpha > 1$, i.e., if one overrelaxes. Though the successive overrelaxation (SOR) scheme has been the method of choice in geophysical problems, it has recently been replaced by the more efficient conjugate gradient and multigrid methods.

Overall, the characteristic of elliptic problems is that the solution must be sought simultaneously at all points of the domain. This implies the inversion of generally large matrices. The main issue is computational efficiency.

B.2.2 Parabolic Problems

Parabolic (and hyperbolic) equations describe time dependent problems. They must be formulated as initial boundary value problems. The prototype parabolic problem is the diffusion equation

$$\partial_t u = D \partial_x \partial_x u \tag{B.22}$$

It is linear and describes the diffusion of a property u in physical space from an initial distribution. In the course of time any gradients are smeared out. Consider a time step Δt such that $t = n\Delta t$, $n = 1, \ldots, N$ and a grid size Δx such that $x = j\Delta x$, $j = 1, \ldots, J$.

The simplest finite difference scheme then is the *explicit* scheme

$$u_j^{n+1} = u_j^n + \hat{D}(u_{j+1}^n - 2u_j^n + u_{j-1}^n) + O(\Delta t, (\Delta x)^2) \tag{B.23}$$

where $\hat{D} = \Delta t D / (\Delta x)^2$. The major concern is the stability of such approximations. Linear stability requires that the spatial Fourier modes $u_j^n = A^n e^{ikj\Delta x}$ do not amplify with time for any wavenumber k. For the explicit finite difference scheme (B.23) this implies

$$A^{n+1} = GA^n \tag{B.24}$$

with amplification factor $G = 1 - 4\hat{D}\sin^2 k\Delta x/2$. The scheme is thus linearly stable if $|G| < 1$ for all k which requires $\hat{D} < 1/2$ or a time step $\Delta t < (\Delta x)^2/2D$. The scheme is conditionally stable.

If we violate this criterion the wave with $k\Delta x/2 = \pi/2$ or of wavelength $2\Delta x$ is the most unstable, i.e., the fastest growing mode. Boundedness for finite times relaxes the stability condition to $|G| \le 1 + O(N^{-1})$. For vector problems G becomes a matrix and the stability criterion is that the spectral radius of the matrix be smaller than one, i.e., all eigenvalues of G must have a magnitude smaller than 1.

The constraint on the maximum allowable time step Δt of the explicit scheme (B.23) can be avoided by using the *implicit* scheme

$$u_j^{n+1} - u_j^n = \hat{D}(u_{j+1}^{n+1} - 2u_j^{n+1} + u_{j-1}^{n+1}) + O(\Delta t, (\Delta x)^2) \qquad (B.25)$$

which requires solving the tridiagonal system $-\hat{D}u_{j-1}^{n+1} + (1 + 2\hat{D})u_j^{n+1} - \hat{D}u_{j+1}^{n+1} = u_j^n$. In this case $G = (1 + 4\hat{D}\sin^2 k\Delta x/2)^{-1}$ which is smaller or equal to one for all k. The scheme is thus unconditionally stable. Arbitrarily large time steps can be taken.

In contrast the *centered* time difference scheme

$$u_j^{n+1} - u_j^{n-1} = 2\hat{D}(u_{j+1}^n - 2u_j^n + u_{j-1}^n) + O((\Delta t)^2, (\Delta x)^2) \qquad (B.26)$$

has $G = \frac{1}{2}[a \pm (a^2 + 4)^{1/2}]$, where $a = 8\hat{D}\sin^2 k\Delta x/2$. Since $|G| \geq 1$ for all $\hat{D} > 0$, the scheme is unconditionally unstable. This scheme was, unfortunately, used by L. F. Richardson in his heroic but failed attempt at weather forecasting in the 1920's (see Sect. 5.1.3).

For the two-dimensional diffusion problem, the explicit scheme leads to $\hat{D} \leq \frac{1}{2}\frac{(\Delta y)^2}{(\Delta x)^2 + (\Delta y)^2}$. For $\Delta x = \Delta y$ the allowable time step is half that of the the one-dimensional problem. For three dimensional problems it is one third of the one-dimensional problems.

Commonly used schemes for parabolic equations are the Crank–Nicholson and the Dufort–Frankel scheme.

B.2.3 Hyperbolic Problems

Hyperbolic equations describe advection or propagation. A disturbance propagates along the characteristics from its point of origin. Its influence is only felt within the wedge between the characteristics. As for parabolic problems solutions can be constructed that march downwind into the "domain of influence". Again stability is the major concern. Physically, stability requires that the algorithm does not propagate information faster than the propagation speeds. This requires a time step $\Delta t \leq \Delta x/c$ or a Courant number $C = c\Delta t/\Delta x \leq 1$. This is the famous Courant-Friedrich-Levy criterion (CFL). Additional issues are numerical diffusion, numerical dispersion, and nonlinear instability.

The simplest hyperbolic problem is the advection equation

$$\partial_t u + c\partial_x u = 0 \qquad (B.27)$$

It has the exact solution $u(x, t) = u_0(x - ct)$ where $u_0(x) = u(x, t = 0)$ is the initial distribution. The initial distribution just propagates along the x-axis without any change of form.

One might attempt to solve this advection equation with an explicit scheme that is forward in time and *centered* in space

$$u_j^{n+1} = u_j^n - \frac{1}{2}C(u_{j+1}^n - u_{j-1}^n) + O(\Delta t, (\Delta x)^2) \qquad (B.28)$$

where C is the Courant number. Linear stability analysis gives the amplification factor $|G| = (1 + C^2 \sin^2 k\Delta x)^{1/2} \geq 1$. This schemes is unconditionally unstable.

In the case $c > 0$ the *upwind* scheme

$$u_j^{n+1} = u_j^n - C(u_j^n - u_{j-1}^n) = (1 - C)u_j^n + Cu_{j-1}^n + O(\Delta t, \Delta x) \qquad (B.29)$$

is stable if $|C| < 1$. However, it is diffusive. This can be seen by rewriting (B.29) in the form

$$u_j^{n+1} = u_j^n - C(u_j^n - u_{j-1}^n) + \frac{1}{2}C(u_{j+1}^n - u_{j+1}^n)$$

$$= u_j^n - \frac{1}{2}C(u_{j+1}^n - u_{j-1}^n) + \frac{1}{2}C(u_{j+1}^n - 2u_j^n + u_{j-1}^n) \qquad (B.30)$$

The first two terms represent the unconditionally unstable advection scheme (B.28). The last term is a second order approximation to a diffusion term $D\partial_x\partial_x u$. The upwind scheme actually solves an advection-diffusion equation with a diffusion coefficient $D = \frac{C\Delta x}{2}$. This diffusion is called numerical diffusion. The unstable scheme for the advection term is stabilized by introducing diffusion.

Numerical dispersion occurs when *second order* schemes are introduced for the spatial coordinates. This can be seen by analyzing the second order scheme

$$\frac{du_j}{dt} + c\frac{u_{j+1} - u_{j-1}}{2\Delta x} = 0 \qquad (B.31)$$

It has solutions $u_j(t) = Ue^{i(kj\Delta x - \omega t)}$, with phase speed $\hat{c}_p = \frac{\omega}{k} = c\frac{\sin k\Delta x}{k\Delta x}$ and group velocity $\hat{c}_g = \frac{d\omega}{dk} = c\cos k\Delta x$. These waves are dispersive (Fig. B.1). Their phase speed depends on the wavenumber, whereas the original advection equation supports nondispersive waves with constant phase speed and group velocity. Any initial distribution thus disperses instead of staying intact. The problem is especially crucial for small scale features since their phase speed approaches zero as $k \to \pi/\Delta x$. Small scale features do not propagate at all but stay stagnant.

These schemes are all not very satisfactory. The first order schemes introduce numerical diffusion that smears out gradients. The second order schemes introduce numerical dispersion. Flux corrective transport (FCT) schemes combine low order diffusive and high order dispersive schemes in a way to minimize diffusive and dispersive distortions.

For the nonlinear advection equation

$$\partial_t u + u\partial_x u = 0 \qquad (B.32)$$

with periodic boundary conditions (i.e, $u(L) = u(0)$) additional issues come into play. One arises from the fact that (B.32) can be rewritten as

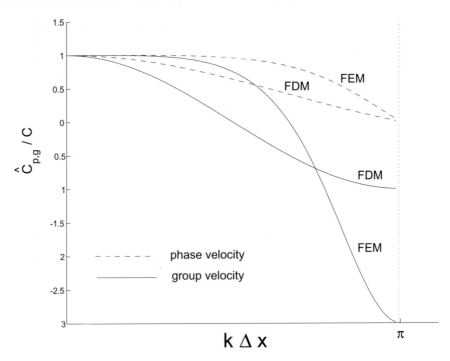

Fig. B.1. The phase speed (*dashed*) and the group velocity (*continuous*) of the finite difference approximation (B.31) (FDM) and of the finite element approximation (B.47) (FEM) of the linear advection equation (B.27)

$$\partial_t u + \partial_x(\frac{1}{2}u^2) = 0 \qquad (B.33)$$

In this form the nonlinear advection equation becomes a conservation equation for a property of density u and flux $F = u^2/2$. The amount of the property within a "volume" only changes due to fluxes across the "surface"

$$\frac{d}{dt}\int_a^b dx u = -[F|_{x=b} - F|_{x=a}] \qquad (B.34)$$

Any numerical algorithm should be designed to not violate this conservation and to not introduce any artificial sources or sinks. This is most easily done by discretizing the flux form (B.33) rather than the advective form (B.32). In the flux form the finite differences are automatically of the form $F_{j+1} - F_j$ and satisfy (B.34) whereas finite differences of the advective form such as $u_j(u_{j+1} - u_j)$ do not necessarily do so. The geophysical equations that represent balance equations for mass, momentum, energy and other properties should thus be discretized in their flux form.

A second issue is that solutions of the nonlinear advection equation may become unstable even if the scheme is linearly stable. Most often variance piles up at small wavelengths $2\Delta x$ and $4\Delta x$, first slowly, then exponentially. The reason for this nonlinear instability is twofold. First, the nonlinearity in the equations couples the Fourier modes. Variance appears in modes that had no variance initially. Second, the smallest resolvable wavelength in a grid is $2\Delta x$. If the nonlinear interactions transfer energy to smaller wavelengths it is folded back or aliased into the larger resolvable wavelengths. This leads to build-up of variance at the smallest wavelengths and eventual instability.

The possibility of this nonlinear instability also expresses itself in the fact that the numerical algorithm does not conserve the same properties as the original differential equation. The advection equation conserves the variance of u or the "kinetic energy" (see (B.34)) for periodic boundary conditions. The finite difference approximation

$$\frac{d}{dt}u_j = -u_j \frac{u_{j+1} - u_{j-1}}{2\Delta x} \tag{B.35}$$

on the other hand gives

$$\frac{d}{dt}\sum_{j=1}^{N} \frac{1}{2}u_j{}^2 = -\frac{1}{2\Delta x}\sum_j (u_j^2 u_{j+1} - u_j^2 u_{j-1})$$

$$= -\frac{1}{2\Delta x}\sum_j (u_j^2 u_{j+1} - u_{j+1}^2 u_j)$$

$$= \frac{1}{2\Delta x}\sum_j u_j u_{j+1}(u_{j+1} - u_j) \tag{B.36}$$

where the right hand side does not necessarily sum to zero.

The same is true for the flux form

$$\frac{d}{dt}u_j = \frac{1}{2}\frac{u_{j+1}^2 - u_{j-1}^2}{2\Delta x} \tag{B.37}$$

which leads to

$$\frac{d}{dt}\sum_{j=1}^{N} \frac{1}{2}u_j{}^2 = -\frac{1}{2\Delta x}\sum_j u_j u_{j+1}(u_{j+1} - u_j) \tag{B.38}$$

and also fails to conserve energy.

The build-up of variance at the smallest wavelengths can be avoided by removing the variance by scale selective filtering or the addition of artificial diffusion (e.g., [142]) or by introducing variance conserving schemes.

If we add *diffusion* to the advection equation we arrive at the advection-diffusion equation

$$\partial_t u + c\partial_x u = D\partial_x \partial_x u \tag{B.39}$$

Consider the scheme

$$u_j^{n+1} = u_j^{n-1} - C(u_{j+1}^n - u_{j-1}^n) + 2\hat{D}(u_{j+1}^{n-1} - 2u_j^{n-1} + u_{j-1}^{n-1}) \qquad \text{(B.40)}$$

If $D = 0$ then the scheme is stable if $|C| \le 1$. If $c = 0$ the scheme is stable if $\hat{D} = D\Delta t/\Delta x^2 < 1/4$. For the general case we find the stability criterion $C^2 \le 1 - 4\hat{D}$, which reduces to the appropriate criterion for the purely advective and diffusive case but implies that the Courant number or time step needs to be reduced when diffusion is added to the advection equation.

B.3 Staggered Grids

The equations describing geophysical flows contain more than one dependent variable. A typical example are the equations

$$\partial_t u - f_o v + g\partial_x \xi = 0$$
$$\partial_t v + f_o u + g\partial_y \xi = 0 \qquad \text{(B.41)}$$
$$\partial_t \xi + h_0(\partial_x u + \partial_y v) = 0$$

which describe inviscid, unforced, linear, barotropic shallow water flow. Here u and v are the horizontal velocity components and ξ the surface elevation; f_0 is the Coriolis parameter, h_0 the constant fluid depth and g the gravitational acceleration. Five different ways to arrange the three dependent variables on a grid are shown in Fig. B.2. They are called the Arakawa A to E grids. The grids A to D have a grid size Δx. The E grid has the same density of points if its grid size is chosen to be $\sqrt{2}\Delta x$.

The different grids perform differently. This can be seen by calculating the dispersion relations of the waves that the grid supports. For $k_y = 0$ we find

$$\left(\frac{\omega}{f_0}\right)^2 = \begin{cases} 1 + \left(\frac{R}{\Delta x}\right)^2 (k\Delta x)^2 & \text{exact} \\ 1 + \left(\frac{R}{\Delta x}\right)^2 \sin^2 k\Delta x & \text{grid } A \\ 1 + 4\left(\frac{R}{\Delta x}\right)^2 \sin^2 k\Delta x/2 & \text{grid } B \\ \cos^2 k\Delta x/2 + 4\left(\frac{R}{\Delta x}\right)^2 \sin^2 k\Delta x/2 & \text{grid } C \\ \cos^2 k\Delta x/2 + \left(\frac{R}{\Delta x}\right)^2 \sin^2 k\Delta x & \text{grid } D \\ 1 + 2\left(\frac{R}{\Delta x}\right)^2 \sin^2 k\Delta x/\sqrt{2} & \text{grid } E \end{cases} \qquad \text{(B.42)}$$

where $R = \sqrt{gh_0}/f_0$ is the Rossby radius of deformation. The exact dispersion relation depends on kR whereas the grid approximations depend on $R/\Delta x$ and $k\Delta x$.

Figure B.3 shows the dispersion relations as a function of $k\Delta x$ for $R/\Delta x = 2$. The dispersion relations are all different and deviate from the true solution. Grids A, D and E show negative group velocities for small wavelengths. For grids B and C zero group velocity is reached at the smallest wavelength $2\Delta x$.

Figure B.4 shows the dispersion relation for three different values of $R/\Delta x$ for the C-grid. For $R/\Delta x > 1/2$ the frequency increases monotonically with

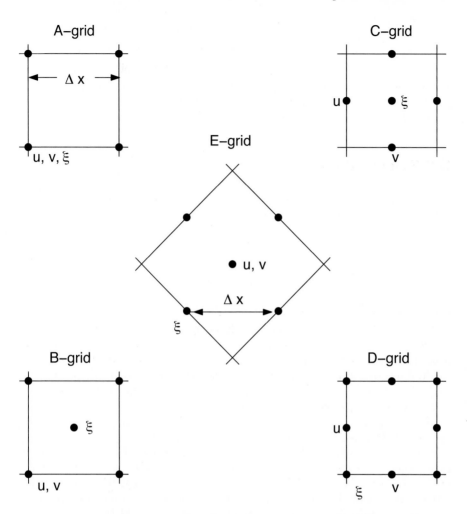

Fig. B.2. Five different ways, A to E, to arrange the three dependent variables u, v and ξ of the shallow water equations on a staggered horizontal grid

k, for $R/\Delta x < 1/2$ it decreases monotonically, and for $R/\Delta x = 1/2$ it is constant. The implications of these differences need to be carefully evaluated when choosing a grid. Most ocean models use the Arakawa B or C grid.

B.4 Spectral Models

Instead of finite difference grids one can also use a truncated set of basis functions to represent the spatial structure. Any function $f(x)$ can be expanded into a generally infinite sum of basis functions $\phi_n(x)$

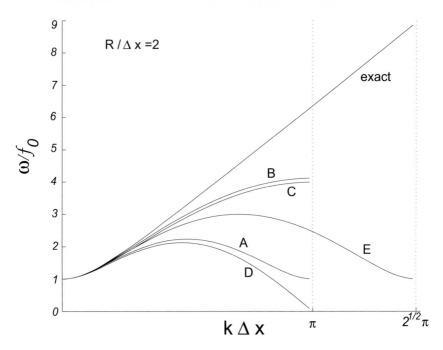

Fig. B.3. Dispersion relations for the exact shallow water equations (B.41) and for the Arakawa A to E grids as a function of $k\Delta x$ for a Rossby radius $R = 2\Delta x$

$$f(x) = \sum_n f_n \phi_n(x) \tag{B.43}$$

Instead of performing operations on the function itself one then has to perform operations on the coefficients f_n and on the basis functions $\phi_n(x)$. Truncation of the expansion at a finite number allows numerical calculations albeit with a truncation error. The most common of these representations for geophysical flows are the spectral and finite element models.

Spectral models represent the latitudinal (φ) and longitudinal (θ) dependence of a function in terms of *spherical harmonics*

$$u(\varphi, \theta, \dots) = \sum_n \sum_m u_n^m(\dots) P_n^m(\mu) e^{im\theta} \tag{B.44}$$

with $\mu = \sin\varphi$ and

$$P_n(\mu) = \frac{1}{2^n n!} \frac{d^n}{d\mu^n} (\mu^2 - 1)^n \tag{B.45}$$

$$P_n^m(\mu) = (1 - \mu^2)^{1/2} \frac{d^m}{d\mu^m} P_n(\mu)$$

and $m = \dots, -2, -1, 0, +1, +2, \dots$ and $n = 1, 2, 3, \dots$ with $|m| \leq n$. Here P_n is the Legendre polynomial of order n and P_n^m the associated Legendre polyno-

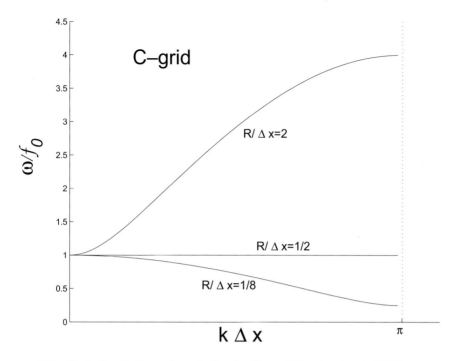

Fig. B.4. C-grid dispersion relation for three different values of $R/\Delta x$

mial of the first kind of degree n and order m. The basis functions $P_n^m(\mu)e^{im\theta}$ are orthogonal and normalized. The index m is the zonal wavenumber. The index $n - |m|$ is the meridional "wavenumber", i.e., the number of zero crossings in the open interval $(-\pi/2, \pi/2)$.

By substituting the expansion into the governing equations, evolution equations for the coefficients $u_n^m(\dots)$ are obtained. The nonlinear advection terms are a major difficulty. They constitute double sums that describe the nonlinear interaction of modes. The number of these interactions is proportional to the square of the number of resolved modes.

The calculation of these nonlinear terms was a major roadblock towards increasing resolution for spectral models. This situation changed in the 1970s when the Fast Fourier Transformation (FFT) became available and B. Machenhauer introduced the transformation method. At each time step, it transforms the nonlinear terms to physical space, evaluates them there and then transforms the result back to spectral wavenumber space, using FFT in both transformations. Gradients are thus evaluated in physical space. Nowadays spectral models have all the variables available in both spectral and physical space at each time step.

Truncation is either done at $N = M$ (triangular truncation) or at $N = |m| + M$ (rhomboidal truncation) (see Fig. B.5). Rhomboidal truncation

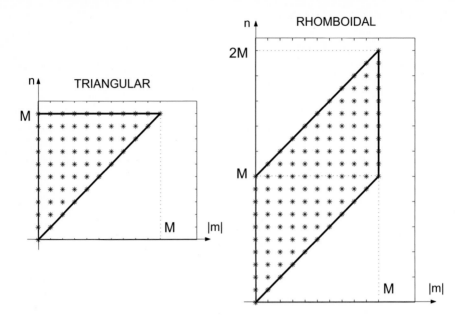

Fig. B.5. Triangular and rhomboidal truncation of spectral models

provides the same meridional resolution for each zonal wavenumber, similar to a regular spatial grid. With only half the number of basis functions, triangular truncation provides the same meridional resolution but only at low zonal wavenumbers.

Spectral models are best suited for the atmosphere, which has no zonal and meridional boundaries. They are ill-suited for the oceans with their complicated basin geometry. When applied to the atmosphere, spectral models offer a couple of advantages. First of all, they represent fields globally. The solution at a point in physical space depends on the solution everywhere else, whereas in a finite difference approximation it only depends on its neighboring points. This global representation conforms more to the underlying physics where "everything is connected with everything else". Also, spectral models can easily accommodate variance conserving schemes.

The major disadvantage of spectral models is inflexibility. Local increase of resolution around mountain ranges or in regions of interest is impossible. Also, they are computationally somewhat less efficient than finite difference models. The number of operations grows at $N \log N$ rather than N, and they are harder to run on massively parallel computers.

B.5 Finite Element Models

Finite elements are another method where the dependent variables are expanded in terms of a set of basis functions. Unlike the global basis functions of spectral methods, the basis functions of the finite element method are local. They are low-order polynomials that are nonzero only in a limited domain. One can thus adopt the grid to any desired resolution locally. This is a great advantage when dealing with inhomogeneous flow fields and complex geometries.

The standard basis function for one-dimensional problems is the "tent" function

$$\phi_j(x) = \begin{cases} 0 & \text{for} \quad x < (j-1)\Delta x \\ [x - (j-1)\Delta x]/\Delta x & \text{for} \quad (j-1)\Delta x \leq x \leq j\Delta x \\ [(j+1)\Delta x - x]/\Delta x & \text{for} \quad j\Delta x \leq x \leq (j+1)\Delta x \\ 0 & \text{for} \quad x > (j+1)\Delta x \end{cases} \qquad \text{(B.46)}$$

Some of the intrinsic advantages and disadvantages of the finite element method can be seen by applying it to the linear advection equation (B.27) which results in

$$\frac{1}{6}\left(\frac{du_{j+1}}{dt} + 4\frac{du_j}{dt} + \frac{du_{j-1}}{dt}\right) + c\frac{(u_{j+1} - u_{j-1})}{2\Delta x} = 0 \qquad \text{(B.47)}$$

Contrary to the finite difference approximation (B.31) the time derivative is smoothed over adjacent grid points.

A wave dispersion analysis gives phase speed $\hat{c}_p = \frac{3c}{k\Delta x} \cdot \frac{\sin k\Delta x}{2+\cos k\Delta x}$ and group velocity $\hat{c}_g = 3c \cdot \frac{1+2\cos k\Delta x}{(2+\cos k\Delta x)^2}$. This phase speed and group velocity is compared in Fig. B.1 with the phase speed and the group velocity obtained for the second order finite difference method (B.31). The dispersion of the finite element method is less than for the finite difference method. However, the $2\Delta x$ wave is still stagnant and has a group velocity $-3c$, three times faster than for the finite different approximation. Small-scale inaccuracies propagate rapidly through the elements and distort the solution.

Finite element methods have successfully been used for two-dimensional barotropic tidal and storm surge models. They were first applied by coastal engineers to fluid problems.

C

Statistical Analysis

Statistical analysis is a difficult concept for the novice. Often it is reduced to the formal determination of relevant parameters, such as mean values, extreme values, auto-covariance functions, spectra or empirical orthogonal functions. Sometimes it is identified with testing hypotheses, by applying a certain formalism and then declaring a result to be "significant"[1]. Statistics indeed encompasses these activities, but essentially statistics is a science which logically accounts for uncertainties when inferring information about a system for which only limited empirical evidence is available. Most often it takes the form of parameter estimation and the determination of the expected uncertainty of these estimates.

Random variable is the key concept in statistical analysis. It is discussed in Sect. C.1. All statistical analyses are related to either characterize random variables with the help of suitable parameters (as discussed in Sect. C.2 "Characteristic Parameters"), or to infer a random variable consistent with limited empirical evidence from data (as discussed in Sect. C.3 "Inference"). Inference is at the core of statistical analysis: to infer rationally from limited empirical evidence, i.e., from a limited number of samples, as much as possible about an assumed, underlying, unknown random variable. This inference is usually done by choosing a method so that on average[2] the estimated parameter is the true value, and that, also on average, the error in estimating this parameter is as small as possible.

In the following we summarize these key elements of statistical analysis. For more details refer to [182] or [188].

---

[1] When a result is found to be "significant", it does not necessarily imply that the result is important or relevant, but that a preconceived concept (or "model" as the statisticians often say) is likely to be *inconsistent* with the data.

[2] "On average" means here that the method is repeatedly applied, and that the outcomes of these repeated applications are averaged.

C.1 Random Variables and Processes

C.1.1 Probability Function

Let us consider a mechanism $\mathbf{X}$ that produces numbers. The outcome is uncertain and differs from realization to realization. Call these numbers $\mathbf{x}_n$ where the index n denotes or counts the different realizations. Assume that all realizations are equivalent and independent, that the uncertainties are the same for each realization and that the value $\mathbf{x}_n$ provides no knowledge about the value of the next realization $\mathbf{x}_{n+1}$ or any other realization. The output is, however, assumed not to be completely irregular, but to satisfy

$$\lim_{N\to\infty} \frac{|\{\mathbf{x}_{n+k} \in [a,b]; k \le N\}|}{N} = F_{\mathbf{X}}([a,b]) \qquad (C.1)$$

for any interval $[a,b]$. The term $|\cdot|$ denotes the number of elements of a set, i.e., in our case the number of realizations with values in the interval $[a,b]$. The function $F_{\mathbf{X}}$ is a non-negative real-valued function, operating on sets of numbers, with the properties

$$F_{\mathbf{X}}([-\infty,+\infty]) = 1$$
$$F_{\mathbf{X}}(\emptyset) = 0 \qquad (C.2)$$
$$F_{\mathbf{X}}([a,b]) + F_x([c,d]) = F_{\mathbf{X}}([a,b] \cup [c,d]) \quad \text{if} \quad b \le c$$
$$F_{\mathbf{X}}([a,b]) \ge 0$$

where $\emptyset$ denotes the empty set. It is called the probability function. The definition (C.1) says that by binning many samples into small intervals we can approximate with increasing accuracy the underlying probability function. The *distribution function* is given by

$$F_{\mathbf{X}}(y) = F_{\mathbf{X}}([-\infty, y]) \qquad (C.3)$$

If the realizations $\mathbf{x}$ can take only discrete values, the distribution function $F_{\mathbf{X}}(y)$ is a step function. When $\mathbf{x}$ varies across a continuum of real numbers, then in most cases the distribution function is differentiable, so that the *probability density function* can be defined as

$$f_{\mathbf{X}}(y) = \frac{dF_{\mathbf{X}}(y)}{dy} \qquad (C.4)$$

The probability of a realization $\mathbf{x}$ having a value within an interval $[a,b]$ is given by

$$p(\mathbf{x} \in [a,b]) = F_{\mathbf{X}}([a,b]) = F_{\mathbf{X}}(b) - F_{\mathbf{X}}(a) = \int_a^b dy\, f_{\mathbf{X}}(y) \qquad (C.5)$$

A mechanism satisfying these conditions is called a *random variable*, named $\mathbf{X}$, with probability distribution function $F_{\mathbf{X}}$ (or probability density function $f_{\mathbf{X}}$).

The distribution function describes the "uncertainty" of the mechanism. The uncertainty is characterized both by the location of the distribution (given for example by the median $\mu = F_{\mathbf{X}}^{-1}(0.5)$) and by the width of the distribution (given for example by the quartile difference $F_{\mathbf{X}}^{-1}(\frac{3}{4}) - F_{\mathbf{X}}^{-1}(\frac{1}{4})$).

For clarity we denote random variables by boldface upper-case letters such as $\mathbf{X}$, their realizations by boldface lower-case letters such as $\mathbf{x}$, and other numbers such as (dummy) arguments of functions by lower-case letters x.

For the above definition of a random variable it is not relevant if the outcome $\mathbf{x}$ of the mechanism $\mathbf{X}$ is *really* random, or if the outcome is deterministic but so complicated that we are unable to disentangle the deterministic rules behind it. If we cannot discriminate the output from that of a random mechanism we may treat it as random. In Fig. 2.1 of Sect. 2.5 we demonstrated that a sum of several deterministic but highly chaotic processes cannot be distinguished from the realization of a white noise process.

There are many families of probability distributions. They include the normal, or Gaussian, the lognormal, the gamma, and the Weibull distributions (see e.g., [188]). Here we briefly discuss the normal, Weibull, and Gumbel distributions.

The most important distribution is clearly the *normal* or *Gaussian distribution*

$$F_n(x) = \frac{1}{2}[1 + \text{erf}\left(\frac{x - \mu}{\sqrt{2}\sigma}\right)]$$

$$f_n(x) = \frac{1}{\sqrt{2\pi}\sigma} \exp\left[-(x - \mu)^2/2\sigma^2\right]$$

where erf is the error function. The parameter μ characterizes the location of the distribution and the parameter σ its width. Its importance is emphasized by the adjective "normal". The reason for its importance is the *Central Limit Theorem*. It states that the distribution of the average of a series of random variables converges towards a normal distribution, when the series becomes longer and longer. The only provisions that need to be satisfied are that the elements in the series are described by the same probability distribution function, and that they are independent from each other (or contain enough independent samples)[3]. The convergence towards "normality" holds for all probability distributions, but for some distributions the convergence is faster (e.g., for symmetric uni-modal ones) than for others (e.g., for highly skewed ones).

In environmental sciences we often consider time mean values, say monthly means or even yearly means. Thus, the considered variable is an average of many equally distributed variables – and in fact, many of the time averaged quantities are nearly normally distributed. The longer the time averaging, the better the normal distribution describes the probability density function. For

---

[3] These conditions may be relaxed. A finite number of distributions that are about equally frequent, suffices, as for instance in case of an annual cycle.

some variables this convergence is fast, as for temperature or nutrient loads, whereas for others, as for rainfall or mixed layer depth, the convergence is less fast.

The Weibull distribution is a 2-parameter distribution given by

$$F_W(x) = 1 - \exp\left[-\left(\frac{x}{\beta}\right)^{\alpha}\right]$$

$$f_W(x) = \frac{\alpha}{\beta}\left(\frac{x}{\beta}\right)^{\alpha-1}\exp\left[-\left(\frac{x}{\beta}\right)^{\alpha}\right]$$

with positive parameters α and β. The parameter α is a shape parameter: probability density functions with $\alpha < 1$ are monotonically decreasing functions with a maximum at $x = 0$; functions with $\alpha = 1$ are identical to exponential distributions; functions with $\alpha \geq 3.6$ are very similar to normal distributions. The parameter β is a scale parameter, i.e., the larger β the broader the distribution.

The Weibull distribution is useful to characterize the distribution of wind speed in the extratropics [6]. As an example, Fig. C.1 shows the histograms (binned frequency distributions) for the observed daily mean wind speed at Ocean Weather Station M derived from 45 years of data, for the four seasons, and Weibull distributions fitted to these histograms. In summer, $\alpha = 2.64$ is smallest and the distribution is most skewed, with a mean wind speed of only 7.3 m/s and an extended tail towards larger values. In winter, $\alpha = 3.04$ is largest and the distribution is almost symmetric, close to normal, and has the strongest mean wind speed.

Among the distributions that describe extreme values the Gumbel distribution or EV-I distribution is the most relevant one. It measures the probability that the maximum of infinitely many samples[4] drawn, say, within a year, is larger than some value x. It is given by

$$F_G(x) = \exp\left\{-\exp\left[-\frac{x-\alpha}{\beta}\right]\right\}$$

$$f_G(x) = \frac{1}{\beta}\cdot\exp\left\{-\exp\left[\frac{x-\alpha}{\beta}\right] - \frac{x-\alpha}{\beta}\right\}$$

Again, it contains two parameters. This time, α represents the location of the maximum of the distribution, and β is again a shape parameter. The distribution is skewed, with the mean value larger than the maximum value. By inverting the distribution function F_G "return values" can be determined by

$$R_T = \alpha - \beta\cdot ln\left[-ln\left(1-\frac{1}{T}\right)\right]$$

On average, once within the time T the threshold R_T will be passed.

---

[4] This is an asymptotic statement. The statement is an approximation for a finite number of samples; how good an approximation depends on the type of distribution from which the samples are drawn.

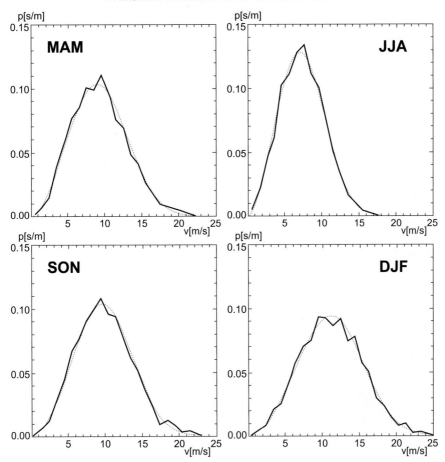

Fig. C.1. Weibull distributions fitted to daily mean wind speeds recorded at the North Atlantic Ocean Weather Station (OWS) M off the Norwegian coast for the four seasons: MAM, JJA, SON und DJF. From Bauer [6]

C.1.2 Bivariate Random Variables

So far we have assumed that **X** generates real numbers, but this assumption can easily be relaxed. In particular complex numbers, vectors, functions or other objects may be generated. Here we first consider bivariate random variables.

A bivariate random variable $\mathbf{Z} = (\mathbf{X}, \mathbf{Y})$ is characterized by a *joint probability density function* $f_{\mathbf{Z}}(x, y)$ which describes the probability that the random variable has a particular outcome $z = (x, y)$. From this joint probability density function one can derive the *marginal density function*

$$f_{\mathbf{X}}(x) = \int_{-\infty}^{\infty} dy \; f_{\mathbf{Z}}(x, y) \qquad (C.6)$$

which describes the probability that the outcome of $\mathbf{X}$ is x, no matter what the outcome of the variable $\mathbf{Y}$. The marginal density function $f_{\mathbf{Y}}(y)$ is defined similarly. The *conditional probability density distribution* is defined by

$$f_{\mathbf{X}|\mathbf{Y}}(x) = \frac{f_{\mathbf{X},\mathbf{Y}}(x, y)}{f_{\mathbf{Y}}(y)} \qquad (C.7)$$

It describes the probability that the outcome of $\mathbf{X}$ is x given that the outcome of $\mathbf{Y}$ is y. The conditional density distribution $f_{\mathbf{Y}|\mathbf{X}}(y)$ is defined likewise. The two variables $\mathbf{X}$ and $\mathbf{Y}$ are called statistically independent when the joint density function has the separable form

$$f_{\mathbf{Z}}(x, y) = f_{\mathbf{X}}(x) f_{\mathbf{Y}}(y) \qquad (C.8)$$

In this case $f_{\mathbf{X}|\mathbf{Y}}(x) = f_{\mathbf{X}}(x)$ and $f_{\mathbf{Y}|\mathbf{X}}(y) = f_{\mathbf{Y}}(y)$. The probability of the outcome x of $\mathbf{X}$ is independent of the outcome y of $\mathbf{Y}$; and the probability of the outcome y of $\mathbf{Y}$ is independent of the outcome x of $\mathbf{X}$. The random variable $\mathbf{X}$ does not provide any information about the variable $\mathbf{Y}$ and vice versa.

An *example* of a bivariate random variable appeared in our discussion of weather forecasting (see Sect. 5.1.5). In this case $\mathbf{X}$ is the forecast, and $\mathbf{Y}$ is the state to be predicted. The top of Fig. 5.15 is an estimate of the joint bivariate distribution. The forecast of a certain weather variable depends very much on the the state to be predicted; otherwise the forecast would be useless. This fact is reflected in the conditional distribution $f_{\mathbf{Y}|\mathbf{X}}(y)$ of the predictand $\mathbf{Y}$ (temperature at some location), given the forecast $\mathbf{X}$. This conditional distribution peaks around $y = x$ and has a narrow width, i.e., when a value x of $\mathbf{X}$ is predicted, then on average the observed temperature y is close to x. The forecast summarized in Fig. 5.15 may therefore be considered useful.

C.1.3 Random Processes

When the output of our mechanism is not numbers or pairs of numbers but is an infinite series $\mathbf{x}_t$ or functions $\mathbf{x}(t)$ one speaks of random processes $\mathbf{X}_t$ or $\mathbf{X}(t)$. The index t is usually taken to be time. It is a discrete index in $\mathbf{X}_t$ with $t = n\Delta t$ and $n = \ldots, -1, 0, 1, \ldots$ and a continuous argument in $\mathbf{X}(t)$. Often $\Delta t = 1$ for convenience. When such random processes are considered, the intervals in equations (C.1, C.2) have to be replaced by different suitable sets and other modifications have to be introduced. However, the overall line of reasoning will not change by these replacements and modifications. Realizations are now infinite series $\mathbf{x}_t$ or functions $\mathbf{x}(t)$. The (joint) probability distribution functions now have these infinite series or functions as their arguments, and are hence quite cumbersome objects. Marginal and conditional distributions can nevertheless be defined in a straightforward manner. Two new important concepts of these random processes are stationarity and memory. To keep the discussion simple we only consider discrete random processes $\mathbf{X}_t$.

A random process $\mathbf{X}_t$ is called stationary if its (joint) probability distribution does not depend on time or, equivalently, is invariant under a time translation. This condition is violated in many applications by trends and cycles, in particular by the annual cycle and the diurnal cycle. The probability for strong winds is larger in winter than in summer. However, in most cases simple transformations produce time series that are approximately stationary. In particular, if $\mathbf{X}_t$ is a process with a trend, then

$$\mathbf{Y}_t = \mathbf{X}_t - a \cdot t \tag{C.9}$$

with an appropriate coefficient a will often be a process that satisfies the stationarity condition. Similarly, when the data include a deterministic cycle of length τ, then any time t may be written as $t = n\tau + t'$ where n is the number of cycles passed so far, and t' the time passed in the present cycle. Then

$$\mathbf{Y}_t = \frac{\mathbf{X}_t - h(t')}{g(t')} \tag{C.10}$$

with two periodic functions $h(t')$ and $g(t')$ will also often produce a more stationary process. In both these transformations it is assumed that the trend or cycle are in the mean. The variability may also have a deterministic trend or cycle. Then additional transformations may yield more stationary time series.

A discrete random process can be viewed as a series of random variables. Each term of the series, $\mathbf{X}_t$ with t fixed, is a random variable. Its probability density function $f_t(x)$ is a marginal density function of the process. Similarly, two terms of the series, $\mathbf{X}_t$ and $\mathbf{X}_{t+k}$ with t and $t + k$ fixed, constitute a bivariate variable. Its joint density function $f_{t,t+k}(x, x')$ can again be viewed as a marginal distribution of the process. For a stationary process it only depends on the time difference or lag k but not on time t itself. For many processes the two variables become independent as the lag k increases. These are processes with a finite memory. The lag needed to obtain independence is an important characteristic of such processes.

The most important random process is the *white noise* process discussed in Sect. 6.1.3. In its discrete version the white noise process $\mathbf{N}_t$ is defined as a process with probability density function

$$f_N(\ldots, x_{t-1}, x_t, x_{t+1}, \ldots) = \prod_{t=-\infty}^{t=\infty} f_n(x_t) \tag{C.11}$$

where $f_n(x)$ is the normal distribution with mean $\mu = 0$ and variance $\sigma^2 = 1$. The white noise process is stationary. All terms in the series are independent from each other. The process has no memory. A continuous version can also be defined. Often the variance is assumed to have an arbitrary value σ_n^2. The importance of the white noise process is that it allows to construct other processes. Prominent examples are *auto-regressive processes*

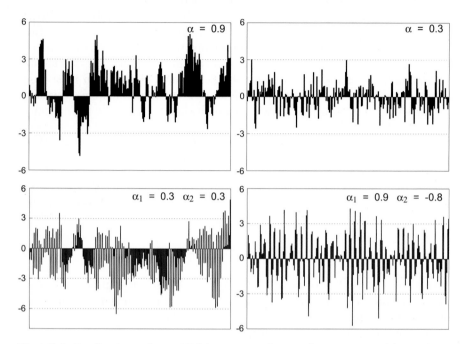

Fig. C.2. Realizations of two AR(1) processes (*top row*) and two AR(2) processes (*bottom row*) for 240 time steps. AR(1), *left:* $a_1 = 0.3$; *right:* $a_1 = 0.9$. AR(2), *left:* $a_1 = a_2 = 0.3$; *right:* $a_1 = 0.9$ and $a_2 = -0.8$; All processes are forced by normally distributed white noise $\mathbf{N}_t$ with zero mean and unit variance. From von Storch and Zwiers [182]

$$\mathbf{X}_t = a_0 + \sum_{k=1}^{K} a_k \mathbf{X}_{t-k} + \mathbf{N}_t \qquad (C.12)$$

They are stationary if $|y_k| > 1$ for all (complex and real) roots y_k of the characteristic polynomial $1 - \sum_k a_k y^k$ [182] (p. 207). K is called the order of the auto-regressive process. An auto-regressive process of order K is often abbreviated AR(K). One can think of (C.12) as a discretized linear ordinary differential equation with constant coefficients and a stochastic forcing term.

The two most important classes of auto-regressive processes are of 1st and 2nd order. A first order process, $\mathbf{X}_t = a_o + a_1 \mathbf{X}_{t-1} + \mathbf{N}_t$ shows a stationary piecewise decaying behavior, while a second order process $\mathbf{X}_t = a_o + a_1 \mathbf{X}_{t-1} + a_2 \mathbf{X}_{t-2} + \mathbf{N}_t$ shows a stationary piecewise decaying or piecewise oscillatory behavior. Figure C.2 shows examples of two first order and two second order AR-processes.

The *stationarity* condition for an AR(1) process is simply $|a_1| < 1$. For an AR(2) process, stationarity requires $a_2 \pm a_1 < 1, |a_2| < 1$.

Stationary AR(1) processes are also called *red noise* processes.

C.2 Characteristic Parameters

Random variables and random processes are fully characterized by their probability density functions. These probability density function are, however, often quite cumbersome, if not impossible, to deal with. Therefore one considers a variety of *parameters* that characterize limited but relevant aspects of a random process (or variable). A parameter can be a number, but it can also be a vector, a matrix or a function. In the first subsection we define the standard parameters: mean, variance, covariance (correlation), auto-covariance function (auto-correlation function) and spectrum. In the subsequent subsections we discuss empirical orthogonal functions, the decomposition of variance, and skill scores, parameters that are specifically employed in atmospheric and oceanic sciences.

C.2.1 Expectation Values

Mean. To define characteristic parameters one needs the concept of *expectation* of a random variable. It is the average of all possible outcomes weighted by their probability density

$$\mu_x = \mathrm{E}(\mathbf{X}) = \int dx \; x \; f_{\mathbf{X}}(x) \qquad (C.13)$$

and is a number. The number μ_x is also called the *mean*. If the distribution is symmetric, half of the outcomes will be larger than μ_x, and half of them smaller.

Variance. If $\mathbf{X}$ is a random variable, then $(\mathbf{X} - \mu_x)^2$ is a random variable as well. It describes the squared distance of an outcome of $\mathbf{X}$ from the mean μ_x. The expectation of this random variable is the *variance*

$$\sigma_x^2 = \mathrm{VAR}(\mathbf{X}) = \mathrm{E}\big((\mathbf{X} - \mu_x)^2\big) \qquad (C.14)$$

where σ_x is called the *standard deviation* and is a measure of the width of the distribution.

The parameters μ and σ in the normal distribution are indeed the mean and standard deviation. For the Weibull distribution the mean and variance are

$$\mu_W = \beta \cdot \Gamma(1 + \frac{1}{\alpha}) \quad \text{and} \quad \sigma_W^2 = \beta^2 \left[\Gamma(1 + \frac{2}{\alpha}) - \Gamma^2(1 + \frac{1}{\alpha}) \right] \qquad (C.15)$$

where $\Gamma(\cdot)$ is the gamma function. For the Gumbel distribution

$$\mu_G = \alpha + 0.57721 \cdot \beta \quad \text{and} \quad \sigma_G^2 = \beta^2 \pi^2 / 6 \tag{C.16}$$

Covariance. When a *bivariate* random variable $\mathbf{Z} = (\mathbf{X}, \mathbf{Y})$ is considered, the strength of the link between the two paired outcomes may be characterized by the *covariance* of the two variables

$$\text{Cov}(\mathbf{X}, \mathbf{Y}) = \text{E}((\mathbf{X} - \mu_x)(\mathbf{Y} - \mu_y)) \tag{C.17}$$

An alternative is the normalized variant of the covariance, the *correlation*

$$\rho_{xy} = \frac{\text{Cov}(\mathbf{X}, \mathbf{Y})}{\sigma_x \sigma_y} \tag{C.18}$$

Covariances vary between $-\infty$ and $+\infty$, and correlations between -1 and $+1$. A zero-value usually indicates, in particular when the joint distribution is normal, that the two random variables are independent. A positive value indicates that we may on average expect outcomes of $\mathbf{X}$ and $\mathbf{Y}$ to be both larger, or both smaller, than their respective means, whereas a negative value indicates a preferences for opposite signs. The correlation is $+1$ if the two variables are linearly related, i.e., if $\mathbf{X} = a + b\mathbf{Y}$.

Auto-covariance function. When a stochastic process is considered, then the link between consecutive outcomes can be measured by the *auto-covariance function*.

$$\gamma_x(\Delta) = \text{Cov}(\mathbf{X}_t, \mathbf{X}_{t+\Delta}) = \text{E}((\mathbf{X}_t - \mu_x)(\mathbf{X}_{t+\Delta} - \mu_x)) \tag{C.19}$$

For stationary processes it only depends on the lag Δ but not on t. Its normalized variant is the *auto-correlation function*

$$\rho_x(\Delta) = \frac{\gamma_x(\Delta)}{\gamma_x(0)} \tag{C.20}$$

For discrete random processes both functions are discrete symmetric functions, i.e., $\Delta = 0, \pm 1, \pm 2, \ldots \pm \infty$ and $\gamma_x(\Delta) = \gamma_x(-\Delta)$. They have a positive maximum at the origin. In case of limited memory processes they level off to zero values after some time. The decay time of this auto-covariance function is often taken as a proxy for the *memory* of the process. Auto-covariance functions of the AR(1) and AR(2)-processes shown in Fig. C.2 are given in [182].

Spectrum. An alternative representation of the time behavior of a process is given by the *spectrum*, which is simply the *Fourier transform* of the auto-covariance function (for more details, see [71] or [182]). The Fourier transform maps the discrete auto-covariance function γ_x to a continuous positive function

$$\Gamma_x(\omega) = \sum_{\Delta=-\infty}^{\infty} \gamma_x(\Delta) e^{-i2\pi\omega\Delta} = \gamma_x(0) + \sum_{\Delta=1}^{\infty} \gamma_x(\Delta) \cos(2\pi\omega\Delta) \tag{C.21}$$

on the *frequency* interval $\omega \in [0, \frac{1}{2}]$. The operation (C.21) is invertible, i.e., the spectrum and the auto-covariance function contain the same information, but they differ with respect to their interpretation.

An important property of the spectrum is that its integral equals the variance

$$\text{VAR}(\mathbf{X}_t) = \gamma_x(0) = 2 \int_0^{\frac{1}{2}} d\omega \, \Gamma_x(\omega) \qquad (C.22)$$

Note that at this point the spectrum has nothing to do with oscillations or harmonic analysis, which decomposes a series of numbers into contributions from different waves or oscillations[5]. A link to harmonic analysis emerges when we address the problem of how to *estimate* a spectrum from a limited time series of numbers $\{\mathbf{x}_0, \ldots, \mathbf{x}_T\}$ in Appendix C.3. The proper interpretation of a spectrum is that the variance $\int_\omega^{\omega+\delta\omega} dx \, \Gamma_x(x)$ in a small frequency band $[\omega, \omega + \delta\omega]$ is due to variations on time scales of ω^{-1}. It does not mean that these variations are in any sense regular oscillations. A *peak* in the spectrum indicates that the considered process varies at that time scale more strongly than at other time scales. When the process is an auto-regressive process, such a peak represents the frequency of an oscillatory eigenmode of that linear process (see [182], chapter 11).

C.2.2 Empirical Orthogonal Functions

Consider a random vector $\mathbf{X} = (\mathbf{X}_1, \ldots, \mathbf{X}_J)^T$ with J components. Then, one can calculate all covariances among the components. This results in the symmetric *covariance matrix*

$$C_x = \begin{pmatrix} \text{VAR}(\mathbf{X}_1) & \text{Cov}(\mathbf{X}_1, \mathbf{X}_2) \ldots & \text{Cov}(\mathbf{X}_1, \mathbf{X}_J) \\ \text{Cov}(\mathbf{X}_2, \mathbf{X}_1) & \text{VAR}(\mathbf{X}_2) & \ldots \text{Cov}(\mathbf{X}_2, \mathbf{X}_J) \\ \vdots & \vdots & \ldots \vdots \\ \text{Cov}(\mathbf{X}_J, \mathbf{X}_1) \text{Cov}(\mathbf{X}_J, \mathbf{X}_2) \ldots & \text{VAR}(\mathbf{X}_J) \end{pmatrix} \qquad (C.23)$$

This matrix describes the co-variability of all components.

C_x is a positive semi-definite matrix. All its eigenvalues λ_j $(j = 1, \ldots, J)$ are real and non-negative, and its eigenvectors $\mathbf{e}^j$ form an orthogonal basis[6].

---

[5] For certain pathological cases such a link can be constructed. Harmonic analysis is also helpful when a deterministic cycle is contained in the data; in this case, however, the autocovariance function does not level off to zero with increasing lag and the Fourier transform (C.21) does not exist.

[6] A number λ is an eigenvalue of a matrix C if there is a non-zero vector $\mathbf{e}$ such that $C\mathbf{e} = \lambda\mathbf{e}$. The vectors $\mathbf{e}$ are not uniquely determined, as they may be multiplied by any constant. However, if the eigenvalues λ are all different, then the directions are uniquely determined. If some of the eigenvalues are identical, then the directions of the associated eigenvectors are arbitrary but can be chosen to be *orthogonal*. Vectors are orthogonal if the scalar product between them vanishes, $\mathbf{e}^{j\,T} \cdot \mathbf{e}^k = 0$ for $j \neq k$.

Therefore the random vector may be expanded into the series

$$\mathbf{X} = \sum_{j=1}^{J} c_j \mathbf{e}^j \tag{C.24}$$

$$c_j = \mathbf{X}^T \cdot \mathbf{e}^j$$

This expansion is simply a coordinate transformation. The state of the system may be given by the components $\mathbf{X}_j$ or by the coefficients c_j. The vectors $\mathbf{e}$ are named *principal vectors* or, particularly in meteorology and oceanography, *empirical orthogonal functions* (EOFs). The coefficients c_j are called *principal components* or *EOF coefficients*.

A similar expansion can be made with respect to any set of orthogonal vectors. The expansion with respect to the eigenvectors $\mathbf{e}$ of the covariance matrix has the property that the truncated expansions

$$\mathbf{X} = \sum_{j=1}^{L} c_j \mathbf{e}^j + \delta_L \tag{C.25}$$

with $L < J$ are more efficient in representing the variance than expansions using any other orthonormal set of vectors. This is true for all truncations "L". "Efficiency" means that the first EOF describes more variance of $\mathbf{X}$ than any other vector, i.e.,

$$\mathrm{VAR}\big(\mathbf{X} - c_1 \mathbf{e}^1\big) = \delta_1^2 = \mathrm{VAR}(\mathbf{X}) - \lambda_1 = \sum_{j=2}^{J} \lambda_j \tag{C.26}$$

is a minimum. The amount of variance left unaccounted for is equal to the sum of all eigenvalues λ_j with $j \geq 2$. Similarly, the EOF $\mathbf{e}^2$ represents the second largest amount of variance of $\mathbf{X}$ that can be described by a vector orthogonal to $\mathbf{e}^1$, and so forth.

In most cases, the size of the eigenvalues is very non-uniform, with a few very large eigenvalues and many very small eigenvalues. In this case, just the first few EOFs are capable of describing the bulk of the variance of the vector $\mathbf{X}$. Thus, the EOFs are efficient in compressing most of the information in a multivariate data set into a smaller-dimensional space. Instead of the full vector $\mathbf{X} = (\mathbf{X}_1, \ldots, \mathbf{X}_J)$ the transformed, truncated vector $\mathbf{X}^L = (c_1, \ldots, c_L)$ with a small number L is used.

EOFs are not necessarily related to certain physical processes. Often this is the case for the first EOF, sometimes for the second, but usually not for the higher-indexed EOFs. Due to construction, all higher-indexed EOFs must be orthogonal to all lower-indexed EOFs, and physical processes are usually not "orthogonal" to each other.

For further reading on EOFs, see the textbooks by Preisendorfer [134], Jolliffe [72] and von Storch and Zwiers [182]. Von Storch and Frankignoul [176] offer a discussion about applications in oceanography.

C.2.3 Decomposition of Variance

The variance of a random variable might be due to different sources. This occurs when one considers a *statistical model*[7] that describes a plausible physical forcing-response link between two variables.

Let us assume that one state variable $\mathbf{Y}$ affects another variable $\mathbf{X}$, like the large scale atmospheric circulation affecting the synoptic variability. Such a link may be conceptualized by assuming that the random variable $\mathbf{X}$ is *conditioned* upon the random variable $\mathbf{Y}$[8]. In this case the probability density function $f_{\mathbf{X}}(x)$ of $\mathbf{X}$ may by partitioned such that

$$f_{\mathbf{X}}(x) = \int dy \; f_{\mathbf{X}|\mathbf{Y}}(x) f_{\mathbf{Y}}(y) \tag{C.27}$$

Here $f_{\mathbf{X}|\mathbf{Y}}$ is the *conditional* probability function of $\mathbf{X}$ provided that the random variable $\mathbf{Y}$ takes the value y, and $f_{\mathbf{Y}}$ is the probability density function of $\mathbf{Y}$. The expectation and the variance of $\mathbf{X}$ may then be written as

$$E(\mathbf{X}) = E(E(\mathbf{X}|\mathbf{Y})_X)_Y \tag{C.28}$$
$$\text{VAR}(\mathbf{X}) = E(\text{VAR}(\mathbf{X}|\mathbf{Y})_X)_Y + \text{VAR}(E(\mathbf{X}|\mathbf{Y})_X)_Y \tag{C.29}$$

where the subscript indicates with respect to which random variable the operation "expectation" and "variance" is to be executed.

Equation C.29 implies that the overall variance can be attributed to two different sources, namely to the mean uncertainty of the conditional distributions, and to the variability of the different conditional means. The variations of the model state may be understood as composed of forced variations, related to $\mathbf{Y}$, and intrinsic variations within the model. The intrinsic variations may be modulated by $\mathbf{Y}$, different values of $\mathbf{Y}$ representing different "regimes", but in many cases this dependence is suppressed and the last term in equation (C.29) assumed to be independent of the value of the forcing $\mathbf{Y}$.

Such a forcing-response model often takes the form of a regression model

$$\mathbf{X} = \mu_0 + \beta\mathbf{Y} + \mathbf{N} \tag{C.30}$$

where $\mathbf{X}$ is the response, $\mathbf{Y}$ the forcing and $\mathbf{N}$ white noise with variance σ_n^2. If the driving process $\mathbf{Y}$ has zero expectation and variance σ_y^2 and is independent of $\mathbf{N}$ then

---

[7] Note that we use the term "model" here to formulate a dynamical link in time or between different state variables. This use is different from its use in mathematical statistics. There a model usually means a set of assumptions about the collection of data and about the probabilistic structure of the problem at hand – such as that the observations are taken from the same random variable, that they are independent, that the probability density function is Gaussian, etc.

[8] Validation of a dynamical model sometimes includes the identification of such links, found in either observational records or in model output.

$$E(\mathbf{X}) = \mu_0$$
$$E(\mathbf{X}|\mathbf{Y})_X = \mu_0 + \beta y$$
$$\text{VAR}(E(\mathbf{X}|\mathbf{Y})_X)_Y = \text{VAR}(\mu_0 + \beta \mathbf{Y}) = \beta^2 \sigma_y^2 \qquad (C.31)$$
$$\text{VAR}(\mathbf{X}|\mathbf{Y})_X = E\big((\mathbf{X} - \mu_0 - \beta \mathbf{Y})^2\big) = \sigma_n^2$$

The resulting decomposition $\sigma_x^2 = \beta^2 \sigma_y^2 + \sigma_n^2$ is a special case of equation (C.29). Part of the $\mathbf{X}$-variance is due to the intrinsic variability (σ_n^2) unrelated to the driving process, and the other part is due to the variability of the driving process (σ_y^2).

C.2.4 Skill Scores

The success of forecasts is described by *skill scores* (see also [100] or [73]). These scores characterize the outcome of the bivariate random variable $(\mathbf{F},\mathbf{P})$ consisting of the forecast $\mathbf{F}$ and the predictand $\mathbf{P}$. The most often used measures are the correlation skill score and the *mean square error*. The correlation between the forecast $\mathbf{F}$ and the predictand $\mathbf{P}$ is the correlation skill score

$$\rho = \frac{\text{COV}(\mathbf{F},\mathbf{P})}{\sqrt{\text{VAR}(\mathbf{F})\text{VAR}(\mathbf{P})}} \qquad (C.32)$$

The correlation skill score is not affected if the forecasts contain a constant bias or if the amplitude of the two differs by a constant factor. The *mean square error* is the expected square error

$$S_{FP}^2 = E\big((\mathbf{F} - \mathbf{P})^2\big) \qquad (C.33)$$

For a perfect forecast, that is, $\mathbf{F} = \mathbf{P}$, the correlation skill score ρ is 1 and the mean square error S_{FP}^2 is zero. If $\mathbf{F}$ is the climatological forecast (i.e., $\mathbf{F} = E(\mathbf{P})$), then $\rho = 0$ and $S_{FP}^2 = \text{VAR}(\mathbf{P})$. If $\mathbf{F}$ is a random forecast, with the same mean and variance as $\mathbf{P}$ then $\rho = 0$ and $S_{FP}^2 = 2\text{VAR}(\mathbf{P})$. Thus, the correlation skill score is constructed so that it has the value 1 for a perfect forecast and zero or less than zero for trivial reference forecasts.

The *proportion of described variance*[9] is the percentage of $\mathbf{P}$-variance that is described by $\mathbf{F}$

$$R_{FP}^2 = \frac{\text{VAR}(\mathbf{P}) - \text{VAR}(\mathbf{F} - \mathbf{P})}{\text{VAR}(\mathbf{P})} = 1 - \frac{\text{VAR}(\mathbf{F} - \mathbf{P})}{\text{VAR}(\mathbf{P})} \qquad (C.34)$$

---

[9] Often, the term "explained" variance is used here. However, this terminology is misleading. According to "Merriam Webster's Collegiate Dictionary" the word "explain" stands for: "1a: to make known. b: to make plain or understandable. 2: to give the reason for or cause of. 3: to show the logical development or relationships of". In the statistical methodology used here, nothing is implying a causal relationship or a dynamical understanding. Thus "described" variance is a more adequate term.

The *Brier skill score* is a measure of the skill of the forecast $\mathbf{F}$ relative to a reference forecast $\mathbf{R}$ of the same predictand $\mathbf{P}$. The comparison is made on the basis of the mean square error of the individual forecasts

$$B_{FRP} = 1 - \frac{S^2_{FP}}{S^2_{RP}} = \frac{S^2_{RP} - S^2_{FP}}{S^2_{RP}} \tag{C.35}$$

The Brier skill score differs from the other scores as it explicitly compares against another forecast. This other forecast is usually a much simpler one, often named a "strawman". If the Brier score is larger than zero, then the (usually more advanced) forecast $\mathbf{F}$ is more skillful than the simpler forecast $\mathbf{R}$. Thus, one would conclude, that the extra complexity required for $\mathbf{F}$ over $\mathbf{R}$ is worth the effort. However, if the Brier skill score is below zero then nothing is gained by using the more complex forecast $\mathbf{F}$.

C.3 Inference

Inference covers two broad areas: the estimation of characteristic parameters from finite samples and the testing of hypotheses. We first discuss the basic aspects of estimation and illustrate them by some standard examples. Then we consider the estimation of auto-correlation functions, spectra and EOFs in some detail, before turning to hypothesis testing.

C.3.1 Basic Aspects of Estimation

Numbers and random variables are different entities. Numbers are realizations of a random variable. As a function of a number is another number so is a function of a random variable another random variable. Thus, the result of the process of manipulating random variables is a random variable as well. We thus have to distinguish between two things: when a mean value of T outcomes $x_1, \ldots x_T$ of a random variable $\mathbf{X}$ is calculated, then a number is calculated, namely the sample mean $\bar{x} = \frac{1}{T} \sum_{j=1}^{T} x_j$; when the *process* of calculating a mean from T samples is considered, then another random variable, namely $\bar{\mathbf{X}} = \frac{1}{T} \sum_{j=1}^{T} \mathbf{X}_j$ is introduced. The sample mean has no uncertainty; it is just a number; the random variable $\bar{\mathbf{X}}$, on the other hand, has uncertainty, since it is a statement about how to do the calculation. The former is a realization of the latter. What is the expectation of the $\bar{\mathbf{X}}$? It is $E(\bar{\mathbf{X}}) = \mu_x$. Thus, calculating the mean of samples is a meaningful way of *estimating* the mean of $\mathbf{X}$. We can even calculate the distribution of the difference between $\bar{\mathbf{X}}$ and μ_x. This difference is again a random variable, with zero mean and a standard deviation of $\sigma_x/\sqrt{T}$ if the samples are drawn independently.

This is an important point, often misunderstood by the novice: When we repeatedly apply the estimation formula $\bar{x} = \frac{1}{T} \sum_{j=1}^{T} x_j$ to many independent samples $x_1, \ldots \ldots x_T$, then the error of this operation $\bar{x} - \mu_x$ will be on average zero, and the standard deviation of this operation will be $\sigma_x/\sqrt{T}$. This is a useful assertion, but it does not imply anything about the error made, when *one* concrete set of samples is used to estimate the mean of the random variable $\mathbf{X}$. The estimation theory makes statements about the accuracy of the *process* of estimating something, not about the error made when a best guess is derived from a sample. When one estimate is calculated, then one realization of a random variable is drawn. This may be a number much larger or much smaller than the mean, and there is no way to find out how close it actually is to the mean value.

This line of argument, presented here with the simplest of all cases, namely the estimation of the mean, applies for *all* estimation problems. When we calculate a *confidence interval* for a parameter, we are not voicing our confidence about the numbers but about the process of calculating these numbers.

In general, an *estimator* $\hat{p}$ of a parameter p of a random variable $\mathbf{X}$ is a function of the random variables $\mathbf{X}_1 \ldots, \mathbf{X}_T$. The estimator is used to infer an estimate or best guess of p by inserting the sample $x_1 \ldots, x_T$ but $\hat{p}$ itself is a random variable. It has a probability density function, an expectation $\mathrm{E}(\hat{p})$ and a variance $\mathrm{VAR}(\hat{p})$. One would like to see the *mean square error* (MSE) small

$$\mathcal{M}(\hat{p}, p) = \mathrm{E}\big((\hat{p} - p)^2\big) \tag{C.36}$$

Generally, estimators make use of all available samples; thus they depend on the sample size T. An estimator is called *consistent* if the MSE converges towards zero for $T \to \infty$. If two estimators for the same parameter p exist, then that one with a smaller MSE is more *efficient*. The mean squared error may be split into two components

$$\mathcal{M}(\hat{p}, p) = [\mathcal{B}(\hat{p})]^2 + \mathrm{VAR}(\hat{p}) \tag{C.37}$$

with the *bias* $\mathcal{B}(\hat{p}) = \mathrm{E}(\hat{p}) - p$ being the expected error of the estimator. The first term on the right hand side of (C.37) represents the systematic error, while the second stands for the irregular fluctuations of the estimator. Often, one wants the bias to be zero – then the estimator is *unbiased*. But the efficiency of an estimator depends on both, the bias *and* the variance.

Example C.1. For the variance σ_x^2 of a random variable, two estimators are in common use, namely $\hat{\sigma}_x^2 = \frac{1}{T}\sum_{j=1}(\mathbf{X}_j - \bar{\mathbf{X}})^2$ and $S^2 = \frac{1}{T-1}\sum_{j=1}(\mathbf{X}_j - \bar{\mathbf{X}})^2$. They differ only with respect to the factor in front of the sum. The former is biased with $\mathcal{B}(\hat{\sigma}_x^2) = \frac{1}{T}\sigma_x^2$ and the latter is unbiased. But the variance of the former is smaller than that of the latter, so that $\mathcal{M}(\hat{\sigma}_x^2, \sigma_x^2) < \mathcal{M}(S^2, \sigma_x^2)$.

There are no strict rules of how to design estimators. Estimators are not right or wrong, but more or less efficient.

One way of constructing efficient estimators is the *Maximum Likelihood method* (ML method). The parameters are determined such that the probability density function takes its maximum value for the observations: Let us assume that we have T realizations $\mathbf{x}_1 \ldots, \mathbf{x}_T$ available to estimate a parameter p. Let $f_{\mathbf{X}}(x, p)$ be the known density function of $\mathbf{X}$ with the unknown parameter p. The joint probability density function of the T independent samples $\mathbf{x}_j$ is then $f_{\mathbf{X}_1 \ldots \mathbf{X}_T}(x_1, \ldots, x_T; p) = \prod_{j=1}^{T} f_{\mathbf{X}}(x_j; p)$. In this formula p is a given parameter and $(x_1 \ldots, x_T)$ the independent arguments. By inserting the realization $\mathbf{x}_1 \ldots, \mathbf{x}_T$ into this formula we obtain the *likelihood function*. In the likelihood function, the realizations are fixed and p is variable. A maximum likelihood estimate of the parameter p is the value of p that maximizes this likelihood function. This estimate depends on the realization. By substituting the random variables for the realizations we obtain an estimator $\hat{p}$, the maximum likelihood estimator. It can be shown under fairly general conditions that the ML method generates consistent and asymptotically optimally efficient estimators. Examples are the sample mean $\bar{\mathbf{X}}$ as an estimator of the mean of a Gaussian random variable $\mathbf{X}$, and $\hat{\sigma}_x^2$ as an estimator of its variance.

When estimating a parameter p one would like to have an interval $[A_L, A_U]$ such that, with a certain probability, the true parameter p is contained in that interval. These intervals are unknown and need to be estimated. One needs estimators $\hat{A}_L$ and $\hat{A}_U$ that are random variables. The interval $[\hat{A}_L, \hat{A}_U]$ is called a *confidence interval*. From it one can calculate the probability $\text{prob}\left(p \in [\hat{A}_L, \hat{A}_U]\right) = q$. Its correct interpretation is that when we repeat the sampling and calculate the interval often enough, then in $q \times 100$ percent of all cases, the (changing) interval will contain the fixed parameter p. Again, nothing is said about any concrete realization. When we calculate a realization $[a_L, a_U]$ of the random variable "confidence interval", then we do not know whether it contains the parameter p or not. The confidence interval does not provide us with confidence about the location of the true parameter, but about the accuracy of the method.

Example C.2. Let $\mathbf{X}$ be a normally distributed variable with mean μ_x and standard deviation σ_x. For simplicity let us assume that σ_x is known. T samples $\mathbf{X}_j$ are available. The considered estimator is that of the mean, $\bar{\mathbf{X}} = \frac{1}{T} \sum_{j=1}^{T} \mathbf{X}_j$, which is unbiased and has the variance σ_x^2/T. To arrive at the confidence band, we transform to the variable $\mathbf{Z} = \sqrt{T}(\bar{\mathbf{X}} - \mu_x)/\sigma_x$ with the unknown μ_x. The distribution of $\mathbf{Z}$ is a standard Gaussian distribution (with zero mean and unit variance). Given q, we can thus determine numbers z_L and $z_U = -z_L$ such that $\text{prob}(\mathbf{z} \in [z_L, z_U]) = q$. Then $\text{prob}\left(\bar{x} - z_U \frac{\sigma_x}{\sqrt{T}} < \mu_x < \bar{x} + z_U \frac{\sigma_x}{\sqrt{T}}\right) = q$, and the q-confidence interval for μ_x is $\left[\bar{\mathbf{X}} - z_U \frac{\sigma_x}{\sqrt{T}}, \bar{\mathbf{X}} + z_U \frac{\sigma_x}{\sqrt{T}}\right]$.

Example C.3. Confidence intervals for correlation coefficients can be obtained by calculating the Fisher transform $\hat{z} = \frac{1}{2} ln \left(\frac{1+\hat{\rho}_{xy}}{1-\hat{\rho}_{xy}}\right)$ of the correlation coefficient estimator $\hat{\rho}_{xy}$. The q-confidence interval for z is approximately given by $\hat{z} \pm Z_{1+q/2}/\sqrt{T-3}$, with $Z_{1+q/2}$ being the $1 + q/2$-percentile of the standard normal distribution. Confidence intervals for other parameters, such as variances and regression coefficients, can be calculated as well.

C.3.2 Estimation of Auto-covariance Functions

Here we consider the estimation of the *auto-correlation function* (C.20). In analogy to the estimator of the variance, one usually uses the estimator

$$\hat{\rho}(\Delta) = \hat{\gamma}(\Delta)/\hat{\gamma}(0) \tag{C.38}$$

with $\hat{\gamma}(\Delta)$ being the sample auto-covariance function estimator

$$\hat{\gamma}(\Delta) = \begin{cases} \frac{1}{T}\sum_{t=1}^{T-\Delta}(\mathbf{X}_t - \bar{\mathbf{X}})(\mathbf{X}_{t+\Delta} - \bar{\mathbf{X}}) & \text{for} \quad 0 < \Delta \leq T - 1 \\ 0 & \text{for} \quad T \leq \Delta \\ \hat{\gamma}(-\Delta) & \text{for} \quad \Delta < 0 \end{cases} \tag{C.39}$$

The estimator (C.38) can have substantial bias. For a white noise process, the bias is $\mathcal{B}\big(\hat{\rho}(\Delta)\big) \approx -\frac{1}{T}$, and for an AR(1) process with coefficient a_1 it is

$$\mathcal{B}(\hat{\rho}(\Delta)) \approx \begin{cases} -\frac{1}{T}(1 + 3a_1) & \text{for} \quad \Delta = 1 \\ -\frac{1}{T}\left(\frac{1+a_1}{1-a_1}(1 - a_1^{|\Delta|}) + 2|\Delta|a_1^{|\Delta|}\right) & \text{for} \quad |\Delta| > 1 \end{cases} \tag{C.40}$$

Under the assumption that $\mathbf{X}_t$ is a stationary normal process, the variability of $\hat{\rho}(\Delta)$ is asymptotically given by

$$\text{VAR}(\hat{\rho}(\Delta)) \approx \frac{1}{T}\sum_{j=-\infty}^{\infty} \left(\rho^2(j) + \rho(j + \Delta)\rho(j - \Delta)\right. \tag{C.41}$$
$$\left. -4\rho(\Delta)\rho(j)\rho(j - \Delta) + 2\rho^2(j)\rho^2(\Delta)\right)$$

Thus, if there exists a p such that $\rho(\Delta)$ is zero for $\Delta \geq p$, then $\text{VAR}(\hat{\rho}(\Delta)) \approx \frac{1}{T}\left(1 + 2\sum_{j=1}^{p}\rho^2(j)\right)$ for $\Delta \geq p$. This result is of importance as it tells us that the estimated auto-correlation function does not decay as quickly as the "true" auto-correlation function . The tail of any estimated auto-correlation function shows significant[10] values, which have no counterpart in the true auto-correlation function. The estimated auto-correlation function may thus lead to the false conclusion that there is memory across very long lags Δ. Even if this cannot be ruled out in certain cases, it appears likely that in almost all cases, these long-term correlations are a mere artifact of the estimation method. In fact, when a long time series is cut into two pieces, then the similarity of the estimate for short time lags Δ is usually high, whereas the tails of the two estimates differ greatly from each other.

Moreover, the estimated auto-correlation function has a complex correlation structure of its own

---

[10] Not meant in the statistical sense!

$$\text{Cov}(\hat{\rho}(\Delta), \hat{\rho}(\Delta + \delta)) \approx \frac{1}{T} \sum_{j=-\infty}^{\infty} \rho(j)\rho(j + \delta)$$

For an AR(1) process with coefficient a_1, this approximation gives a correlation of $\hat{\rho}(\Delta)$ and $\hat{\rho}(\Delta + \delta)$ of approximately a_1^{δ} at large lags Δ. That is, the correlations between the auto-correlation function estimates are roughly similar to those of the process itself. When these correlations are persistent, then the estimated auto-correlation function will vary slowly around zero even when the real auto-correlation function has dropped off to zero. This is another reason for exercising even more care when interpreting the tail of the estimated auto-correlation functions.

For further details refer to [71] and [182].

C.3.3 Estimation of Spectra

The estimation of the auto-covariance function has many inherent problems. *Estimating spectra* and assessing the uncertainty of these estimates is even more demanding.

There are two main approaches to estimate the *spectrum* (C.21). One approach consists in making a harmonic analysis of the sample $x_1 \dots, x_T$

$$x_t = a_0 + \sum_{j=1}^{q} a_j \cos(2\pi\omega_j t) + b_j \sin(2\pi\omega_j t) \tag{C.42}$$

with expansion coefficients a_j and b_j, frequencies $\omega_j = j/T$ and $q = T/2$. The sample size T has been assumed to be even. The *periodogram* is then defined by

$$I_{Tj} = \frac{T}{4}(a_j^2 + b_j^2) \tag{C.43}$$

It distributes the sample variance among the different frequencies:

$$\text{VAR}(X_t) = \frac{2}{T} \sum_{j=1}^{q-1} I_{Tj} + \frac{1}{T} I_{Tq}$$

The periodogram provides asymptotically unbiased estimates of the spectrum at the frequencies $\omega_j = 0, 1/T, 2/T, \dots 1/2$.

The alternative is to use the definition of the spectrum (C.21) and to calculate the Fourier transform of the estimated auto-covariance function:

$$\hat{\Gamma}_x(\omega) = \hat{\gamma}_x(0) + \sum_{\Delta=1}^{T} \hat{\gamma}_x(\Delta)\cos(2\pi\omega\Delta) \tag{C.44}$$

Different from the periodogram, this is an estimate of the continuous spectrum[11], ranging from $[0, \frac{1}{2}]$. It returns the same numbers as the first estimate at the frequencies $\omega_j = 0, 1/T \dots, \frac{T-1}{2}/T, 1/2$.

---

[11] This is an artifact due to the fact that the estimator (C.39) of the autocovariance function is extended to infinite lags.

Neither estimator is consistent. When the time series gets longer, the estimation errors do not get smaller; instead the number of frequencies increases. Furthermore, the estimates suffer from severe variability. The periodogram is approximately distributed as

$$\begin{array}{ll} \Gamma(0) \cdot \chi^2(1) & \text{for} \quad \omega_j = 0 \\ \Gamma(\tfrac{1}{2}) \cdot \chi^2(1) & \text{for} \quad \omega_j = \tfrac{1}{2} \\ \tfrac{1}{2}\Gamma(\omega_j) \cdot \chi^2(2) & \text{for all other} \quad \omega_j \end{array} \qquad (\text{C.45})$$

if T is even. Here $\chi^2(k)$ represents the chi-square distribution with k degrees of freedom. A relevant detail is that the periodogram at different frequencies varies independently, at least asymptotically.

Thus, both estimates are more or less useless, even if they appear appealing to the physically trained novice. The reason is that we are trying to estimate a "parameter" of a random process. The sample $x_1 \ldots, x_T$ represents T samples of the random variable X_t with t fixed but only one sample of the random process X_t with $t = \ldots, -1, 0, 1, \ldots$

The situation can be saved, though, in various ways. One way is to use an extension of the time series to split the time series into "chunks" and to calculate a periodogram for each chunk separately (Bartlett chunk method). One then has more than one sample of the random process and the different periodograms can be averaged, which leads to a significant reduction of the variance of the estimate (proportional to the number of chunks). This splitting into chunks can be made in such a manner that both the number of frequencies is increased and the variance reduced, so that the estimator becomes consistent. This method can also be applied to the Fourier-transform of the estimated auto-covariance function.

The alternative is to calculate the full periodogram (or the Fourier-transform of the full estimated auto-covariance function) and to smooth the estimates from neighboring resolved frequencies. This method is usually more efficient – in terms of reducing variability – than the Bartlett chunk method. A variety of smoothing functions are available, which go under names such as Daniell, Parzen and Bartlett. For further details refer to [71] and [182].

When the data are drawn from an auto-regressive process, then spectra may be estimated by fitting an autoregressive process to the data, and by using the spectrum of the estimated process as an estimate for the spectrum. Any application of this method, of course, *presumes* that the considered process is closely approximated by an AR process of suitable order.

C.3.4 Estimation of EOFs

The estimation of *empirical orthogonal functions* (EOFs) is done straightforwardly by first calculating the sample covariance matrix and then computing the eigenvalues and eigenvectors of this positive semi-definite matrix. When the number T of independent samples is less than the length K of the random vector $\mathbf{X}$, then only T non-zero eigenvalues can be found, and the remaining $K - T$ eigenvalues are zero.

Not much can be said about the uncertainty of estimated EOFs (see [72]). There are some rules available about the expected error of the eigenvalues, but hardly anything for the vectors themselves. However, it is a general experience that the vectors associated with the largest eigenvalues are usually robustly estimated, whereas the higher indexed EOFs usually exhibit large variability[12]. Thus, in cases when details of the vectors matter, and not just their efficiency in compressing data into a few degrees of freedom, the practitioner is advised to resort to methods of *resampling* and splitting. The full data sample is divided into independent subsamples, EOFs and eigenvalues are calculated from the subsamples. Then the robustness of the estimated vectors may be determined by comparing EOFs derived from different subsamples. Also the efficiency in compressing data may be tested by calculating the vectors from one subsample, and by determining their efficiency from another subsample.

There is one group of techniques that allegedly leads to the identification of "significant EOFs". However, this terminology is a misnomer. These techniques deal with a rather special problem. When eigenvalues are identical then the directions of the associated eigenvectors are no longer uniquely determined. The eigenvalues are "degenerate". When two eigenvalues are the same, then *all* vectors from a two-dimensional linear subspace qualify as eigenvectors.

In such a situation, the estimation process will bring forward very different vectors when different samples are used, not so much because of sampling variability, but because of the inherent degeneracy of the vectors. Therefore some rules have been devised to identify eigenvalues that may be equal. Experience has shown that, in fact, most of the high-indexed EOFs of high-dimensional random vectors have eigenvalues that cannot be distinguished from a series of small but equal eigenvalues; thus the eigenvectors belonging to the tail of the eigenspectrum may all be degenerate and should not be interpreted physically. They nevertheless may be useful in compressing data in an efficient manner.

C.3.5 Hypothesis Testing

Another widely used inference approach is the *test of a hypothesis*. The basic idea is to define a "reference", and to examine whether a system is consistent with this reference. To do so, empirical evidence about the system is gathered and compared with the expected statistics of the reference. If this comparison leads to the assessment that the evidence is unlikely to emerge from the reference, the *null hypothesis* "system complies with reference" is *rejected*. If the

---

[12] It is actually the spacing between the eigenvalues that matters. A small spacing implies large variability. In most environmental applications, the largest eigenvalues are well separated, while the higher indexed eigenvalues are often very close together.

assessment finds no contradiction then the null hypothesis is not rejected, but it is also not accepted. Instead a weaker statement is made, namely that the evidence is not inconsistent with the reference. It may be that at a later time, when more evidence has been gathered, the null hypothesis will be rejected.

To introduce the concept more formally, we consider a simple prototypical case. Let $\mathbf{X}$ be random variable with density function $f_{\mathbf{X}}$. This is the statistical model, upon which the test is based. We have one realization $\mathbf{x}'$, but it is unknown if $\mathbf{x}'$ has been drawn from $\mathbf{X}$ or not. The null hypothesis, usually denoted by H_0, is "$\mathbf{x}'$ is drawn from $\mathbf{X}$". We determine the smallest possible range Θ of outcomes of $\mathbf{X}$ so that $\mathrm{prob}(\mathbf{x} \in \Theta) = \tilde{p}$ with some pre-specified, normally large probability $\tilde{p}$. When $\mathbf{X}$ is a univariate Gaussian distribution with zero mean and density function f_n, then $\Theta = [-d, d]$ with $\int_{-d}^{d} dy\, f_n(y) = \tilde{p}$. If the sample $\mathbf{x}'$ lies in the interval, we consider $\mathbf{x}'$ as consistent with $\mathbf{X}$, even if we admit that it may be drawn from a distribution very similar to that of $\mathbf{X}$.

However, if $|\mathbf{x}'| > d$, then we consider this a sufficiently unlikely event under the null hypothesis – and we *decide* to reject the null hypothesis. When $\mathbf{X}$ is a bivariate Gaussian distribution, the "region of non-rejection" Θ is an ellipse (given by the covariance matrix) so that the integral of the probability density function over this area is just $\tilde{p}$. The situation is sketched for both the uni- and bivariate cases in Fig. C.3. The null hypothesis is rejected for $\mathbf{x}'$, and it is not for $\mathbf{x}''$.

The above line of argument is objective apart from the choice of $\tilde{p}$. The number $1-\tilde{p}$ is called the *significance level* and gives the acceptable probability to erroneously reject the null hypothesis. It must be selected subjectively. Its choice has something to do with the risk of a false rejection that one is willing to accept. Obviously, this choice depends on the context within which the decision is made. Instead of a risk of $1 - \tilde{p}$, also the term "testing at a significance level of $1 - \tilde{p}$" is used. When the null hypothesis is rejected, the finding is declared "significant at the $1 - \tilde{p}$ level"[13]. The probability of correctly rejecting the null hypothesis, the *power*, is not known. It is at least as large as the significance level and may be arbitrarily close to the significance level, since the correct model to describe the outcome $\mathbf{x}'$ may be very close to the tested model $\mathbf{X}$. Generally, the concept of "power" is theoretically useful, but the power can in most practical situation not be determined because it depends on the unknown true parameter. However, the power of all reasonable tests increases with increasing sample size, for a given deviation from the null hypothesis.

---

[13] Note that the use of the term *significance* is different here from its colloquial use, where it represents "the quality of being important".

Even the statistical term "significance" is often used in a confusing manner, as for instance in: "95% significant" when $\tilde{p} = 95\%$, or "the null hypothesis is rejected with a 95% confidence level". Correct language would be that the result is "significant at the 5% level" or "the null hypothesis is rejected with a 5% risk of error".

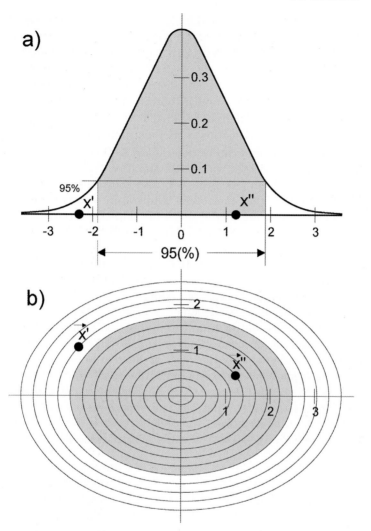

Fig. C.3. Schematic diagrams illustrating the univariate (*top*) and bivariate (*bottom*) domains Θ for which the null hypothesis "**x** is drawn from **X**" is not rejected. The points **x**′ are examples of samples that provide evidence contrary to the null hypothesis, whereas the realizations **x**″ are consistent with the null hypothesis. From von Storch and Zwiers [182]

The most widely used test of a null hypothesis is the *t-test*. It addresses the problem whether a sample $\mathbf{x}_1, \ldots \mathbf{x}_T$ is consistent with the assumption that the unknown random variable $\mathbf{X}$, which has generated the samples, has an expected value of μ_0. Formally: $H_0 : \mathrm{E}(\mathbf{X}) = \mu_0$ with some number μ_0. The model is adopted that $\mathbf{X}$ is a normally distributed random variable, of which neither the mean μ_x nor the standard deviation σ_x are known. If the null hypothesis is true, then the derived variable $\mathbf{Y} = \sqrt{T}\frac{\bar{\mathbf{X}}-\mu_0}{S}$ with $\bar{\mathbf{X}} = \frac{1}{T}\sum \mathbf{X}_j$ and the sample standard deviation S, is t-distributed with $T - 1$ degrees of freedom. Again, a number d can be determined so that $\int_{-d}^{d} dy\, f_t(y) = \tilde{p}$, and the null hypothesis is rejected with a risk of $\tilde{p}$ if $|\mathbf{y}| > d$.

The *detection* problem (see Sect. 5.4) also constitutes a hypothesis test. The null hypothesis is that the recent temperature trends, or other appropriately defined events, are in the range of "normal" variations. When the trend is found to be at the outer fringes of the distribution, then this does not imply that the trend is *not* caused by internal climate variations but only that it is highly unlikely; the alternative explanation – that the trend is due to anthropogenic causes – is considerably more plausible.

Model validation is another case of a hypothesis test. The model simulation is compared with observations or analyses. The null hypothesis is that the statistics of the model and of the observations are identical. The test can be formulated in terms of mean values and/or of variances. When numerical experiments are made to study the effect of different boundary conditions or different parameterizations, then one simulation with "control" and one with "experimental" conditions is done. The comparison of the two again takes the form of a statistical test, using the null hypothesis of equal statistics. In the first case, the validation case, the desired result is to not reject the null hypothesis, whereas in the second case, the sensitivity analysis, it is rejection.

D

Data Assimilation

Data assimilation (DA) refers to techniques that combine models and data for a variety of purposes. DA techniques consists of a set of observations (or other data), a dynamical model and a blending scheme that combines the two. Most of these techniques fall into two categories: filtering and smoothing. Filtering is the sequential update of the state vector to improve the forecast. Smoothing combines model output and data to obtain optimal field estimates. These two techniques can be put into a common framework which is summarized in this appendix, following [141]. For simplicity of presentation it is assumed that both the model and observation equations are linear. The basic methodology of DA is first demonstrated by a simple example in Sect. D.1. Then filtering is discussed in Sect. D.2. It uses the optimal Kalman filter as a framework to describe other less optimal filters. Two smoothing methods, the adjoint and inverse method are discussed in Sect. D.3. These smoothing methods can easily be generalized to include parameter estimation and to arrive at optimal sampling strategies. More details on DA techniques in meteorology and oceanography can be found in [8], [22], [43], [76], [141] and [191].

Data assimilation is a relatively new approach within environmental sciences; however, it receives more and more attention as it helps to overcome one of the fundamental problems in environmental sciences, namely that the segments of the environment cannot be observed in their entirety. The greatest success of DA is its operational use in weather forecasting, and it is expected that similar routines will become common in other fields as well, as our examples in Chaps. 5 and 6 illustrate.

D.1 Estimation

Consider two estimates $\hat{X}_1$ and $\hat{X}_2$ of a random variable X with

$$\begin{aligned}
\mathrm{E}\left(\hat{X}_1\right) &= \mathrm{E}\left(\hat{X}_2\right) = \mathrm{E}(X) \\
\mathrm{VAR}\left(\hat{X}_i\right) &= \sigma_i{}^2 \qquad \text{for} \quad i = 1, 2 \\
\mathrm{Cov}\left(\hat{X}_1, \hat{X}_2\right) &= 0
\end{aligned} \tag{D.1}$$

The estimates are unbiased, have variances $\sigma_1{}^2$ and $\sigma_2{}^2$ and are uncorrelated. What is the best estimate that can be constructed from $\hat{X}_1$ and $\hat{X}_2$? Try the linear combination

$$\hat{X}_a = a_1 \hat{X}_1 + a_2 \hat{X}_2 \tag{D.2}$$

where the subscript a stands for "analysis". For this estimate to be unbiased $a_1 + a_2 = 1$ must be fulfilled or $\hat{X}_a = \hat{X}_1 + a_2(\hat{X}_2 - \hat{X}_1)$. The variance of this estimate is $\mathrm{VAR}\left(\hat{X}\right) = (1 - a_2)^2 \sigma_1{}^2 + a_2^2 \sigma_2{}^2$. The minimum occurs when

$$a_2 = K = \frac{\sigma_2{}^{-2}}{\sigma_1{}^{-2} + \sigma_2{}^{-2}} \tag{D.3}$$

The optimal estimate therefore is

$$\hat{X}_a = \hat{X}_1 + K(\hat{X}_2 - \hat{X}_1) \tag{D.4}$$

with inverse variance

$$\sigma_a{}^{-2} = \sigma_1{}^{-2} + \sigma_2{}^{-2} \tag{D.5}$$

If we interpret the subscript 1 as model and the subscript 2 as data we have all the ingredients of a data assimilation scheme: The two estimates $\hat{X}_1$ and $\hat{X}_2$ correspond to the dynamical and observational model. In order to proceed one needs to specify the errors of these two models. The linear combination (D.2) of the two estimates is the blending scheme. The optimal estimate (D.4) is obtained by minimizing the variance of the blended estimate. The variance of the blended estimate is calculated in (D.5). The DA schemes discussed in the next sections are application of this procedure to specific dynamical and observational models.

D.2 Filtering

The discussion of filtering schemes is best started with the Kalman filter. The Kalman filter is the optimal DA scheme for a certain class of filtering problems. It provides a convenient framework to describe other less optimal schemes.

D.2.1 Kalman Filter

The Kalman filter is based on the following assumptions, definitions and derivations:

1. The dynamical and observational models are assumed to be linear. The model and observation equations (3.2) and (3.3) thus take the form

$$\boldsymbol{\psi}_{i+1} = \mathbf{A}_i \boldsymbol{\psi}_i + \boldsymbol{\epsilon}_i \tag{D.6}$$

and

$$\boldsymbol{\omega}_i = \mathbf{C}_i \boldsymbol{\psi}_i + \boldsymbol{\delta}_i \tag{D.7}$$

where $\mathbf{A}_i$ is a $M \times M$ matrix and $\mathbf{C}_i$ a $M \times N$ matrix. M is the dimension of the state vector, N the dimension of the observation vector and i the discrete time index.

2. The errors in the dynamical and observational models are assumed to have the following properties:

$$
\begin{aligned}
\mathrm{E}(\boldsymbol{\epsilon}_i) &= \mathrm{E}(\boldsymbol{\delta}_i) &&= 0 &&\text{for } i = 0 \ldots K \\
\mathrm{E}\left(\boldsymbol{\epsilon}_i \boldsymbol{\epsilon}_j^T\right) &= \mathrm{E}\left(\boldsymbol{\delta}_i \boldsymbol{\delta}_j^T\right) &&= 0 &&\text{for } i, j = 0, \ldots, K; \ i \neq j \\
\mathrm{E}\left(\boldsymbol{\epsilon}_i \boldsymbol{\epsilon}_i^T\right) &&&= \mathbf{Q}_i &&\text{for } i = 0 \ldots K \\
\mathrm{E}\left(\boldsymbol{\delta}_i \boldsymbol{\delta}_i^T\right) &&&= \mathbf{R}_i &&\text{for } i = 0 \ldots K \\
\mathrm{E}\left(\boldsymbol{\epsilon}_i \boldsymbol{\delta}_j^T\right) &&&= 0 &&\text{for } i, j = 0 \ldots K
\end{aligned}
\tag{D.8}
$$

where $\mathbf{Q}_i$ is the $M \times M$ covariance matrix of the dynamical error and $\mathbf{R}_i$ the $N \times N$ covariance matrix of the observational error. Usually both error matrices are assumed to be independent of the time index i. The dynamical and observational errors are uncorrelated. With these specifications the dynamical model becomes a multivariate autoregressive process.

3. For any estimate $\hat{\boldsymbol{\psi}}_i$ at time i an estimate at time $i+1$ can be obtained by

$$\hat{\boldsymbol{\psi}}_{i+1} = \mathbf{A}_i \hat{\boldsymbol{\psi}}_i \tag{D.9}$$

The covariance matrix of this estimate

$$\mathbf{P}_i = \mathrm{E}\left((\hat{\boldsymbol{\psi}}_i - \boldsymbol{\psi}_i)(\hat{\boldsymbol{\psi}}_i^T - \boldsymbol{\psi}_i^T)\right) \tag{D.10}$$

satisfies the recursion formula

$$\mathbf{P}_{i+1} = \mathbf{A}_i \mathbf{P}_i \mathbf{A}_i^T + \mathbf{Q}_i \tag{D.11}$$

since $\hat{\boldsymbol{\psi}}_i - \boldsymbol{\psi}_i$ and $\boldsymbol{\epsilon}_i$ are uncorrelated.

4. The blending scheme is assumed to be the linear combination

$$\hat{\boldsymbol{\psi}}_i^a = \hat{\boldsymbol{\psi}}_i + \mathbf{K}_i[\boldsymbol{\omega}_i - \mathbf{C}_i \hat{\boldsymbol{\psi}}_i] \tag{D.12}$$

where $\mathbf{K}_i$ is the $N \times M$ gain matrix and $(\boldsymbol{\omega}_i - \mathbf{C}_i \hat{\boldsymbol{\psi}}_i)$ the data-model misfit. The error covariance matrix of this estimate is $\mathbf{P}_i^a = \mathbf{P}_i - \mathbf{P}_i \mathbf{C}_i^T \mathbf{K}_i^T - \mathbf{K}_i \mathbf{C}_i \mathbf{P}_i + \mathbf{K}_i (\mathbf{C}_i \mathbf{P}_i \mathbf{C}_i^T + \mathbf{R}_i) \mathbf{K}_i^T$ because $(\hat{\boldsymbol{\psi}}_i - \boldsymbol{\psi}_i)$ and $\boldsymbol{\delta}_i$ are uncorrelated. It becomes minimal for

$$\mathbf{K}_i = \mathbf{P}_i \mathbf{C}_i^T [\mathbf{R}_i + \mathbf{C}_i \mathbf{P}_i \mathbf{C}_i^T]^{-1} \tag{D.13}$$

and then has the value

$$\mathbf{P}_i^a = \mathbf{P}_i - \mathbf{K}_i \mathbf{C}_i \mathbf{P}_i \tag{D.14}$$

Kalman filtering then consists of the following steps:

- Specify initial conditions $\hat{\boldsymbol{\psi}}_0^a$ and their error covariance matrix $\mathbf{P}_0^a$
- Obtain forecast

$$\hat{\boldsymbol{\psi}}_{i+1} = \mathbf{A}_i \hat{\boldsymbol{\psi}}_i^a \tag{D.15}$$

- Obtain forecast error

$$\mathbf{P}_{i+1} = \mathbf{A}_i \mathbf{P}_i^a \mathbf{A}_i^T + \mathbf{Q}_i \tag{D.16}$$

- Compute Kalman gain

$$\mathbf{K}_{i+1} = \mathbf{P}_{i+1} \mathbf{C}_{i+1}^T [\mathbf{R}_{i+1} + \mathbf{C}_{i+1} \mathbf{P}_{i+1} \mathbf{C}_{i+1}^T]^{-1} \tag{D.17}$$

- Update forecast

$$\hat{\boldsymbol{\psi}}_{i+1}^a = \hat{\boldsymbol{\psi}}_{i+1} + \mathbf{K}_{i+1} [\boldsymbol{\omega}_{i+1} - \mathbf{C}_{i+1} \hat{\boldsymbol{\psi}}_{i+1}] \tag{D.18}$$

- Update error covariance matrix

$$\mathbf{P}_{i+1}^a = \mathbf{P}_{i+1} - \mathbf{K}_{i+1} \mathbf{C}_{i+1} \mathbf{P}_{i+1} \tag{D.19}$$

Kalman filtering is thus a recursive sequential procedure. It not only provides estimates $\hat{\boldsymbol{\psi}}_i^a$ for $i = 1, \ldots, K$ given an initial condition $\hat{\boldsymbol{\psi}}_0^a$ but also the error covariance matrix $\mathbf{P}_i^a$ of these estimates given the covariance matrix of the initial condition $\mathbf{P}_0^a$, and $\mathbf{R}_i$ and $\mathbf{Q}_i$. The Kalman filter is the optimal sequential filter for linear systems under the stated assumptions. Though optimal it has a number of deficiencies:

(i)   The Kalman filter is the optimal filter only if the dynamical and observational models are linear. While one can always linearize more general models the time step of the Kalman filter is given by the time bins into which observations have been collated. This time step is usually much larger than any time interval for which the dynamical model can reasonably be linearized.

(ii) The Kalman filter depends crucially on the model covariance matrices $\mathbf{R}_i$ and $\mathbf{Q}_i$ which are generally poorly known.

(iii)   The calculation of the error covariance matrix $\mathbf{P}_i^a$ is computationally very demanding. It requires M^4 multiplications per time step and the storage of M^2 variables. For GCMs with several state variables and 10^6 grid points this is computationally not feasible with present day computers.

For these reasons Kalman filtering is only applied to models with low dimensional state vectors. For high dimensional state vectors less optimal but computationally feasible filtering schemes have been developed. These schemes are generally of a more heuristic nature as can be seen by the following examples.

D.2.2 Optimal or Statistical Interpolation

Optimal or statistical interpolation (OI) uses the Kalman filtering scheme but prescribes the filter gain and does not calculate the error covariance matrices. OI thus uses the recursion formula

$$\hat{\psi}_{i+1} = \mathbf{A}_i \hat{\psi}_i^a \tag{D.20}$$

and the blending scheme

$$\hat{\psi}_{i+1}^a = \hat{\psi}_{i+1} + \mathbf{K}_{i+1} \left[\omega_{i+1} - \mathbf{C}_{i+1} \hat{\psi}_{i+1} \right] \tag{D.21}$$

where the filter gain $\mathbf{K}_{i+1}$ is prescribed. OI is currently the most popular DA scheme for numerical weather prediction. The gain is usually determined from the grid to measurement point correlation matrix and the measurement error covariance matrix.

D.2.3 Nudging

Nudging of point observations ω was originally introduced as a non-statistical method to relax the dynamical model to observations. In continuous notation this is achieved by adding a relaxation term $(\omega - \psi)/\tau$ to the governing dynamical equation

$$\frac{d\psi}{dt} = A\psi + \frac{\omega - \psi}{\tau} \tag{D.22}$$

Here τ is the relaxation time that needs to be prescribed. The smaller the relaxation time the closer does the dynamical model follow the observations. The value of the relaxation time should thus depend on the model and the data uncertainties. It must be smaller than the decorrelation time over which initial conditions affect the evolution of the dynamical system. For a multidimensional state vector equation (D.22) may be written

$$\frac{d\psi}{dt} = A\psi + \mathbf{K}\mathbf{C}^{-1}(\omega - \mathbf{C}\psi) \tag{D.23}$$

where $\mathbf{C}^{-1}$ is the generalized inverse of $\mathbf{C}$ and $\mathbf{K}$ a diagonal $M \times M$ matrix of relaxation rates. Discretizing this equation in time leads to

$$\psi_i = \mathbf{A}_{i-1}\psi_{i-1} + \mathbf{K}_i\mathbf{C}_i^{-1}(\omega_i - \mathbf{C}_i\mathbf{A}_{i-1}\psi_{i-1}) \qquad (\text{D.24})$$

which shows that nudging can also be regarded as a Kalman filtering scheme with prescribed gain matrix.

The nudging technique described in the reconstruction of regional weather in Sect. 5.2.2 is an example where nudging is applied in the spectral domain (cf. Appendix B.4).

D.2.4 Blending and Direct Insertion

Blending consists of the recursion formula

$$\hat{\psi}_i = \mathbf{A}_{i-1}\hat{\psi}_{i-1}^a \qquad (\text{D.25})$$

and the blending scheme

$$\psi_i^a = a\omega_i + (1 - a)\mathbf{C}_i\hat{\psi}_i \qquad (\text{D.26})$$

at grid points where data are available and

$$\psi_i^a = \hat{\psi}_i \qquad (\text{D.27})$$

at the other grid points. The parameter a with $0 \le a \le 1$ needs to be prescribed. It determines how much weight is given to the data, a, and how much weight is given to the model, $(1 - a)$, and should be inferred from the model and data uncertainties. Direct insertion is the special case $a = 1$. In this case one assumes that observations do not have any errors.

D.2.5 Minimization

As discussed in Sect. 3.2 the various filtering schemes can also be obtained by simultaneously minimizing the "distance"

$$C_i = \parallel \omega_i - \mathbf{C}_i\psi_i^a \parallel + \parallel \psi_i^a - \psi_i \parallel \qquad (\text{D.28})$$

between data and analysis and between analysis and model (see also Fig. 3.3). For quadratic norms $\mathbf{N}_i^\circ$ and $\mathbf{N}_i^m$ this distance becomes $C_i = (\omega_i^T - \psi_i^{aT}\mathbf{C}_i^T) \mathbf{N}_i^\circ(\omega_i - \mathbf{C}_i\psi_i^a) + (\psi_i^{aT} - \psi_i^T)\mathbf{N}_i^m(\psi_i^a - \psi_i)$ and minimization leads to

$$\psi_i^a = \psi_i + \mathbf{K}_i[\omega_i - \mathbf{C}_i\psi_i] \qquad (\text{D.29})$$

with gain

$$\mathbf{K}_i = [\mathbf{N}_i^m + \mathbf{C}_i^T\mathbf{N}_i^\circ\mathbf{C}_i]^{-1}\mathbf{C}_i^T\mathbf{N}_i^\circ \qquad (\text{D.30})$$

Different choices of $\mathbf{N}_i^\circ$ and $\mathbf{N}_i^m$ lead to different filtering schemes. The choice $\mathbf{N}_i^\circ = \mathbf{R}_i^{-1}$ and $\mathbf{N}_i^m = \mathbf{P}_i^{-1}$ reproduces the Kalman gain (D.13). The minimization of a distance or costfunction is also the method of choice for the smoothing problem, as discussed next.

D.3 Smoothing

Smoothing methods (see also Fig. 3.4) construct an optimal field estimate $\psi_i, i = 1, \ldots, K$ from a complete data time series $\omega_i, i = 1, \ldots, K$ and a dynamical model. This optimal field estimate is obtained by minimizing the model-data misfit, weighted by some cost (or penalty or risk) function. The minimization employs the calculus of variation. Smoothing methods have their roots in control theory. They differ in the choice of the control variables which are the variables that are allowed to vary in order to obtain the minimum. The standard methods designate the field estimates themselves or their initial conditions as control variables, but the methods can easily be modified to include parameters, boundary conditions, or forcing fields as control parameters.

The cost function must be specified by the user. It is an a priori and subjective choice which depends on the goal of the study. Typically one employs a weighted sum of the squares of the model-data misfit with the weights being inversely proportional to error covariances. The more accurate the model or data are the more weight they get. In the following subsections we discuss the adjoint method which treats the dynamical model as a strong constraint, the inverse method which treats the dynamical model as a weak constraint, and parameter estimation.

D.3.1 Adjoint Method

The adjoint method regards the dynamical model as a strong constraint. The model is assumed error-free. The final field estimate is a solution of the dynamical equations. The free parameters or control parameters are the initial conditions.

In this case the cost function is the sum of two contributions. One penalizes the uncertainty of the initial conditions $\psi_0 - \psi_0^{(0)}$, where $\psi_0^{(0)}$ is an initial guess of the initial condition. This uncertainty is weighted by the inverse of the error covariance matrix $\mathbf{P}_0$ of the initial conditions. The second contribution penalizes the model-data misfit $\boldsymbol{\delta}_i = \boldsymbol{\omega}_i - \mathbf{C}_i \psi_i$ at all times $i = 1 \ldots K$ weighted by the inverse of the data error covariance matrix $\mathbf{R}_i$ and summed over all i. The strong constraint that $\psi_i = \mathbf{A}_{i-1} \psi_{i-1}$ for all time $i = 1 \ldots K$ is added to the cost function with undetermined Lagrange multipliers λ_i. The cost function to be minimized then becomes

$$C = (\psi_0 - \psi_0^{(0)})^T \mathbf{P}_0^{-1} (\psi_0 - \psi_0^{(0)}) + \sum_{i=1}^{K} 2\boldsymbol{\lambda}_{i-1}^T (\psi_i - \mathbf{A}_{i-1}\psi_{i-1})$$

$$+ \sum_{i=1}^{K-1} (\boldsymbol{\omega}_i^T - \psi_i^T \mathbf{C}_i^T) \mathbf{R}_i^{-1} (\boldsymbol{\omega}_i - \mathbf{C}_i \psi_i) \tag{D.31}$$

where we have used the convention that there are no data at time $i = 0$ and $i = K$. Setting to zero the derivatives with respect to the $2K + 1$ free variables

$\boldsymbol{\lambda}_{i-1}$ for $i = 1, \ldots, K$, $\boldsymbol{\psi}_0$, $\boldsymbol{\psi}_i$ for $i = 1, \ldots, K - 1$ and $\boldsymbol{\psi}_K$ one obtains the $2K + 1$ equations

$$\boldsymbol{\psi}_i = \mathbf{A}_{i-1}\boldsymbol{\psi}_{i-1}, \quad \text{for} \quad i = 1, \ldots, K \tag{D.32}$$

$$\boldsymbol{\psi}_0 = \boldsymbol{\psi}_0^{(0)} + \mathbf{P}_0 \mathbf{A}_0^T \boldsymbol{\lambda}_0 \tag{D.33}$$

$$\boldsymbol{\lambda}_{i-1} = \mathbf{A}_i^T \boldsymbol{\lambda}_i + \mathbf{C}_i^T \mathbf{R}_i^{-1}(\boldsymbol{\omega}_i - \mathbf{C}_i \boldsymbol{\psi}_i), \text{ for } i = 1, \ldots, K - 1 \tag{D.34}$$

$$\boldsymbol{\lambda}_{K-1} = 0 \tag{D.35}$$

The field estimates $\boldsymbol{\psi}_i$ are obtained by forward recursion of (D.32), with initial condition $\boldsymbol{\psi}_0$ (D.33). The Lagrange multipliers $\boldsymbol{\lambda}_i$ are obtained by backward recursion of equation (D.34) with initial condition $\boldsymbol{\lambda}_{K-1} = 0$. This backward recursion is governed by the adjoint operator $\mathbf{A}_i^T$. The forward recursion requires $\boldsymbol{\lambda}_0$. The backward recursion requires $\boldsymbol{\psi}_{K-1}$. The equations thus constitute a two-point boundary value problem, which are notoriously hard to solve. There exists software that calculates the adjoint model of any discrete forward model [44].

D.3.2 Inverse Method

Inverse methods assume that the dynamical model is only a weak constraint. The model is assumed to have errors. The final field estimate is not a solution of the dynamical model. One thus minimizes the cost function

$$\begin{aligned} C = &\ (\boldsymbol{\psi}_0 - \boldsymbol{\psi}_0^{(0)})^T \mathbf{P}_0^{-1}(\boldsymbol{\psi}_0 - \boldsymbol{\psi}_0^{(0)}) \\ &+ \sum_{i=1}^{K-1} (\boldsymbol{\omega}_i^T - \boldsymbol{\psi}_i^T \mathbf{C}_i^T) \mathbf{R}_i^{-1}(\boldsymbol{\omega}_i - \mathbf{C}_i \boldsymbol{\psi}_i) \\ &+ \sum_{i=1}^{K} (\boldsymbol{\psi}_i^T - \boldsymbol{\psi}_{i-1}^T \mathbf{A}_{i-1}^T) \mathbf{Q}_{i-1}^{-1}(\boldsymbol{\psi}_i - \mathbf{A}_{i-1}\boldsymbol{\psi}_{i-1}) \end{aligned} \tag{D.36}$$

Setting to zero the derivatives with respect to the $K + 1$ free variables $\boldsymbol{\psi}_i$ for $i = 0, \ldots, K$ one obtains the equations

$$\boldsymbol{\psi}_i = \mathbf{A}_{i-1}\boldsymbol{\psi}_{i-1} + \mathbf{Q}_{i-1}\boldsymbol{\lambda}_{i-1}, \quad \text{for} \quad i = 1, \ldots, K \tag{D.37}$$

$$\boldsymbol{\psi}_0 = \boldsymbol{\psi}_0^{(0)} + \mathbf{P}_0 \mathbf{A}_0^T \boldsymbol{\lambda}_0$$

if one introduces the variables

$$\boldsymbol{\lambda}_{i-1} = \mathbf{A}_i^T \boldsymbol{\lambda}_i + \mathbf{C}_i^T \mathbf{R}_i^{-1}(\boldsymbol{\omega}_i - \mathbf{C}_i \boldsymbol{\psi}_i), \text{ for } i = 1, \ldots, K - 1$$

$$\boldsymbol{\lambda}_{K-1} = 0 \tag{D.38}$$

These equations are very similar to those for the adjoint method, except that the dynamical model equation contains the error term $\mathbf{Q}_{i-1}\boldsymbol{\lambda}_{i-1}$. This error term also couples the forward and backward recursions and complicates numerical solutions even more.

D.3.3 Parameter Estimation

If the matrix $\mathbf{A}$ depends on parameters $\boldsymbol{\alpha} = (\alpha_1, \ldots, \alpha_L)$ these can also be estimated by including $\boldsymbol{\alpha}$ in the list of control parameters. This is done by

- introducing the parameter equation $\boldsymbol{\alpha} = \boldsymbol{\alpha}^{(0)} + \boldsymbol{\gamma}$ where $\boldsymbol{\alpha}^{(0)}$ is the initial guess of the parameter and $\boldsymbol{\gamma}$ its associated error with zero mean and covariance matrix $\mathbf{S}$, and
- adding a term $\delta C = (\boldsymbol{\alpha} - \boldsymbol{\alpha}^{(0)})^T \mathbf{S}^{-1} (\boldsymbol{\alpha} - \boldsymbol{\alpha}^{(0)})$ to the cost function.

Differentiating this extended cost function then also with respect to $\boldsymbol{\alpha}$ yields the additional equations from which the optimal parameter values can be determined. This approach usually treats the dynamical model and the initial conditions as weak constraints. The model and initial conditions can also be treated as strong constraints. Then the model and initial condition are added to the cost function with associated undetermined Lagrange multipliers, as in equation (D.31). This strong constraint approach often leads to physically and mathematically ill-posed problems since the parameter $\boldsymbol{\alpha}$ might attempt to correct for errors that are really in the dynamical model or initial condition, and a poor estimate of $\boldsymbol{\alpha}$ may result.

The actual application of any of the above data assimilation schemes to a specific problem requires modifications. These modifications must account for forcing and boundary conditions, for dynamical equations that are perhaps not of the autoregressive type, and for any peculiarities of the problem at hand. They are often technically quite challenging. Some of these modifications can be seen in the examples of Chaps. 5 and 6.

References

1. Arakawa A (1988) Finite-difference methods in climate modeling. In: Physically-based modeling and simulation of climate and climatic change, part 1. Kluwer Academic Publishers
2. Arakawa A, Lamb VR (1977) Computational design of the basic dynamical processes of the UCLA general circulation model. In: Methods in computational physics 17:173-265
3. Arrhenius SA (1896) On the influence of carbonic acid in the air upon the temperature of the ground. Phil Mag J Sci 41:237-276
4. Backhaus JO (1993) Das Wetter im Meer: Nord- und Ostsee. Klimaänderung Küste A. 1/5338:37-49
5. Bakan S, Chlond A, Cubasch U, Feichter J, Graf H, Grassl H, Hasselmann K, Kirchner I, Latif M, Roeckner E, Sausen R, Schlese U, Schriever D, Schult I, Schumann U, Sielmann F, Welke W (1991) Climate response to smoke from the burning oil wells in Kuwait. Nature 351:367-371
6. Bauer E (1996) Characteristic frequency distributions of remotely sensed in situ and modelled wind speeds. Int J Climatol 16:1087-1102
7. Bengtsson L (1999) From short-range barotropic modelling to extended-range global weather prediction: A 40-year perspective. Tellus 51A-B:13-32
8. Bennett AF (1992) Inverse methods in physical oceanogrpahy. Cambridge Monographs on Mechanics and Applied Mathematics. Cambridge University Press, Cambridge
9. Bogdanov K, Magarik V (1967) Numerical solutions to the problem of distributions of semidiurnal tides M_2 and S_2 in the world ocean (transl.), Dokl Akad Nauk SSSR 172:1315-1317
10. Bray D, Krück C (2001) Some patterns of interaction between science and policy. Germany and climate change. Clim Res 19:69-90
11. Bray D, von Storch H (1999) Climate science. An empirical example of post-normal science. Bull Am Met Soc 80:439-456
12. Brettschneider G (1967) Modelluntersuchungen der Gezeiten der Nordsee unter Anwendung des hydrodynamisch-numerischen Verfahrens. Mitt Inst Meereskd Univ Hamb 8, Hamburg
13. Brückner E (1890) Klimaschwankungen seit 1700 nebst Bemerkungen über die Klimaschwankungen der Diluvialzeit. Geographische Abhandlungen herausgegeben von Prof. Dr. Albrecht Penck in Wien; Wien and Olmütz, ED

Hölzel (English anthology with partial translations, see Stehr N, von Storch H (eds) (2000) Eduard Brückner – The sources and consequences of climate change and climate variability in historical times. Kluwer Publisher).

14. Brückner E (1886) Weather prophets [Wetterpropheten]. Jahresbericht der Berner Geographischen Gesellschaft

15. Bryan F (1986) High-latitude salinity effects and interhemispheric thermohaline circulations. Nature 323:301-304

16. Bunde A, Havlin S (eds) (1994) Fractals in science. Springer, Berlin Heidelberg New York

17. Charney JG, Fjörtoft R, Neumann J v (1950) Numerical integration of the barotropic vorticity equation. Tellus 2:237-254

18. Charney J, DeVore JG (1979) Multiple flow equilibria in the atmosphere and blocking. J Atmos Sci 36:1205-1216

19. Cotton WR, Pielke RA (1992) Human impacts on weather and climate. ASTeR Press Ft Collins

20. Crowley TJ, North GR (1991) Paleoclimatology. Oxford University Press, New York

21. Cubasch U, Hasselmann K, Höcht H, Maier-Reimer E, Mikolajewicz U, Santer BD, Sausen R (1992) Time-dependent greenhouse warming computations with a coupled ocean–atmosphere model. Clim Dyn 8:55-69

22. Daley R (1991) Atmospheric data analysis. Cambridge University Press, Cambridge.

23. Davies HC (1976) A lateral boundary formulation for multi-level prediction models. Quart J Roy Met Soc 102:405-418

24. De Cosmo J, Katsaros KB, Smith SD, Anderson RJ, Oost WA, Bumke K, Chadwick H (1996) Air–sea exchange of water vapour and sensible heat: The Humidity EXchange Over the Sea (HEXOS) results. J Geophys Res 101,C5: 12,001-12,016

25. Defant, A (1953) Ebbe und Flut des Meeres, der Atmosphäre und der Erdfeste. Springer, Berlin Göttingen Heidelberg

26. Denis B, Laprise R, Caya D, Ct J (2002) Downscaling ability of one-way-nested regional climate models: The big-brother experiment. Clim Dyn 18:627-646

27. Desai SD, Wahr JM (1995) Empirical ocean tide models estimated from TOPEX/POSEIDON altimetry. J Geophys Res 100:25205-25228

28. Dick S, Kleine E, Müller-Navarra SH, Klein H, Komo H (2001) The operational circulation model of BSH (BSHcmod) – Model description and validation. Berichte des Bundesamtes für Seeschifffahrt und Hydrographie 29

29. Dobrovolski SG (2000) Stochastic climate theory and its application. Springer, Berlin Heidelberg New York

30. Egbert GD, Bennett AF, Foreman MGG (1994) TOPEX/POSEIDON tides estimated using a global inverse model. J Geophys Res 99:24821-24852

31. Egbert GD, Ray RD (2001) Estimates of M_2 tidal energy dissipation from TOPEX/POSEIDON altimeter data. J Geophys Res 106:22475-22502

32. Egbert GD, Ray RD (2000) Significant dissipation of tidal energy in the deep ocean inferred from satellite altimeter data. Nature 45:775-778

33. Egger (1988) Alpine lee cyclogenesis: Verification of theories. J Atmos Sci 45:2187-2203

34. Ekman VW (1905) On the influence of the earth's rotation on ocean-currents. Reprinted from Arkiv för matematik, astronomi, och fysik, 2,11, 1963

35. Exner FM (1908) Über eine erste Annäherung zur Vorausberechnung synoptischer Wetterkarten. Meteorol Z 21:1-7

36. Feser F, Weisse R, von Storch H (2001) Multidecadal atmospheric modelling for Europe yields multi-purpose data. EOS 82:305+310

37. Flather RA (2000) Existing operational oceanography. Coastal Engineering 41:1-3,13-40

38. Fleck L (1935) Entstehung und Entwicklung einer wissenschaftlichen Tatsache: Einführung in die Lehre vom Denkstil und Denkkollektiv. Benno Schwabe und Co. (Reprinted 1980 in Suhrkamp Verlag, Frankfurt am Main)

39. Frankignoul C (1995) Climate spectra and stochastic climate models. In: von Storch H, Navarra A (eds) Analysis of climate variability: Applications of statistical techniques. Springer, Berlin Heidelberg New York

40. Frankignoul C (1985) Sea surface temperature anomalies, planetary waves, and air–sea feedback in the middle latitudes. Rev Geophys 23:357-390

41. Friedman RM (1989) Appropriating the weather. Vilhelm Bjerknes and the construction of a modern meteorology. Cornell University Press

42. Funtowicz SO, Ravetz JR (1985) Three types of risk assessment: A methodological analysis. In: Whipple C, Covello VT (eds) Risk analysis in the private sector. Plenum, New York

43. Ghil M, Malanotte-Rizzoli P (1991) Data assimilation in meteorology and oceanography. Advances in geophysics, vol 33. Academic Press, San Diego, CA

44. Giering R, Kaminski T (1996) Recipies for adjoint code construction. Tech Report 212, Max-Planck Institut für Meteorologie, Hamburg

45. Gill AE (1982) Atmosphere–ocean dynamics. Academic Press, international geophysics series 30

46. Giorgi F, Mearns LO (1991) Approaches to the simulation of regional climate change: a review. Rev Geophys 29:191-216

47. Giorgi F, Whetton PH, Jones RG, Christensen JH, Mearns LO, Hewitson B, von Storch H, Fransico R Jack C (2001) Emerging patterns of simulated regional climatic changes for the 21st century due to anthropogenic forcings. Geophys Res Lett 28,17:3317-3320

48. Glahn HR, Lowry DA (1972) The use of model output statistics (MOS) in objective weather forecasting. J Appl Meteor 11:1203-1211

49. Grossmann W (2001) Entwicklungsstrategien in der Informationsgesellschaft – Mensch, Wirtschaft und Umwelt. Springer, Berlin Heidelberg New York

50. Hagner C (2000) European regulations to reduce lead emissions from automobiles – did they have an economic impact on the german gasoline and automobile markets? Reg Env Change 1:135-151

51. Haidvogel DB, Beckmann A (1999) Numerical ocean circulation modelling. Imperial college press, series on environmental science and management

52. Hansen AR, Sutera A (1986) On the probability density function of planetary scale atmospheric wave amplitude. J Atmos Sci 43:3250-3265

53. Hansen W (1952) Gezeiten und Gezeitenströme der halbtägigen Hauptmondtide M_2 in der Nordsee. Dt Hydr Zeitschr, Erg-Heft 1

54. Haltiner GJ, Williams RT (1980) Numerical prediction and dynamic meteorology, second ed. John Wiley and Sons

55. Hasselmann K (1997) Multi-pattern fingerprint method for detection and attribution of climate change. Clim Dyn 13:601-612

56. Hasselmann K (1993) Optimal fingerprints for the detection of time dependent climate change. J Clim 6:1957-1971
57. Hasselmann K (1988) PIPs and POPs: The reduction of complex dynamical systems using principal interaction and oscillation patterns. J Geophys Res 93:11015-11021
58. Hasselmann K (1976) Stochastic climate models. Part I: Theory. Tellus 28:473-485
59. Hegerl GC, Hasselmann KH, Cubasch U, Mitchell JFB, Roeckner E, Voss R, Waszkewitz J (1997) Multi-fingerprint detection and attribution analysis of greenhouse gas, greenhouse gas-plus-aerosol and solar forced climate change. Clim Dyn 13:613-634
60. Held IM, Hou AY (1980) Nonlinear axially symmetric circulations in a nearly inviscid atmosphere. J Atmos Sci 37:515-533
61. Hesse MB (1970) Models and analogies in science. University of Notre Dame Press, Notre Dame
62. Hildebrandson HH (1897) Quelque recherches sure les centres d' action de l' atmosphère. K Sven Vetenskaps akad Handl 29:1-33
63. Holton JR (1972) An introduction to dynamic meteorology. Academic Press
64. Honerkamp J (1994) Stochastic dynamical systems: Concepts, numerical methods, data analysis. VCH, New York
65. Houghton JT, Meira Filho LG, Callander BA, Harris N, Kattenberg A, Maskell K (eds) (1996) Climate change 1995. The science of climate change. Cambridge University Press
66. Huntington E (1925) Civilization and climate. Yale University Press, New Heaven, 2nd ed
67. Huntington E, Visher SS (1922) Climatic changes. Yale University Press, New Heaven
68. Jacob D, Podzun R (1997) Sensitivity studies with the regional climate model REMO. Meteorol Atmos Phys 63:119-129
69. James PM, Fraedrich K, James IA (1994) Wave-zonal-flow interaction and ultra-low-frequency variability in a simplified global circulation model. Quart J Roy Met Soc 120:1045-1067
70. Janssen F (2002) Statistische Analyse mehrjähriger Variabilität der Hydrographie in Nord- und Ostsee. PhD thesis, University of Hamburg
71. Jenkins GM, Watts DG (1968) Spectral analysis and its application. Holden-Day
72. Jollife IT (2002) Principal Component Analysis. Second ed. Springer, Berlin Heidelberg New York
73. Jollife IT, Stephenson DB (eds) (2003) Forecast verification: A practitioner's guide in atmospheric science. Wiley, Chichester
74. Kalnay E, Kanamitsu M, Baker WE (1990) Global numerical weather prediction at the National Meteorological Center. Bull Am Met Soc 71:1410-1428
75. Kalnay E, Kanamitsu M, Kistler R, Collins W, Deaven D, Gandin L, Iredell M, Saha S, White G, Woollen J, Zhu Y, Chelliah M, Ebisuzaki W, Higgins W, Janowiak J, Mo KC, Ropelewski C, Wang J, Leetmaa A, Reynolds R, Jenne R, Joseph D (1996) The NCEP/NCAR 40-Year reanalysis project. Bull Am Met Soc 77:437-471
76. Kantha LH, Clayson CA (2000) Numerical models of the oceans and oceanic processes. Academic Press

77. Kantha LH, Tierney C, Lopez JW, Desai SD, Parke ME, Drexler L (1995) Barotropic tides in the global oceans from a nonlinear tidal model assimilaiting altimetric tides 2. Altimetric and geophysical implications. J Geophys Res 100:25309-25317

78. Kauker F (1999) Regionalization of climate model results for the North Sea. PhD thesis, University of Hamburg

79. Kauker F, von Storch H (2000) Statistics of synoptic circulation weather in the North Sea as derived from a multi-annual OGCM simulation. J Phys Oceano 30:3039-3049

80. Kempton W, Boster JS, Hartley JA (1995) Environmental values in american culture. MIT Press, Cambridge MA and London

81. Kempton W, Craig PP (1993) European perspectives on global climate change. Environment 35:16-45

82. Kirtman BP, Shukla J, Balmaseda M, Graham N, Penland C, Xue Y, Zebiak S (2001) Current status of ENSO forecast skill. A report to the CLIVAR working group on seasonal to interannual prediction. WCRP Informal Report No 23/01 ICPO Publication 56

83. Kirtman BP, Shukla J, Huang B, Zhu Z, Schneider EK (1997) Multiseasonal predictions with a coupled tropical ocean–global atmosphere system. Mon Wea Rev 125:789-808

84. Kistler R, Kalnay E, Collins W, Saha S, White G, Woollen J, Chelliah M, Ebisuzaki W, Kanamitsu M, Kousky V, van den Dool H, Jenne R, Fiorino M (2001) The NCEP/NCAR 50-year reanalysis. Bull Am Met Soc 82:247-267

85. Klein WH, Glahn HR (1974) Forecasting local weather by means of model output statistics. Bull Am Met Soc 55:1217-1227

86. Klinenberg E (2002) Heat Wave. A social autopsy of disaster in Chicago. The University of Chicago Press

87. Kolmogorov AN (1941) The local structure of turbulence in an incompressible viscous fluid for very large Reynolds number. C R Acad Sci, USSR, 30:301-305

88. Köppen V (1923) Die Klimate der Erde. De Gruyter, Berlin

89. Koslowski G, Glaser R (1999) Variations in reconstructed ice winter severity in the Western Baltic from 1501 to 1995, and their implications for the North Atlantic Oscillation. Clim Change 41:175-191

90. Kuhn TS (1970) The structure of scientific revolutions, 2nd ed. University of Chicago

91. Lakoff G, Johnson M (1980) Metaphors we live by. The University of Chicago Press, Chicago London

92. Lamb HH (1979) Climatic variation and changes in the wind and ocean circulation: The Little Ice Age in the northeast Atlantic. Quatern Res 11:1-20

93. Lamb HH (1997) Through the changing scenes of life. A meteorologist's tale. Taverner Publication

94. Landsea CW, Knaff J (2000) How much skill was there in forecasting the very strong 1997-98 El Niño? Bull Am Met Soc 81:2107-2119

95. Langenberg H, Pfizenmayer A, von Storch H, Sündermann J (1999) Storm related sea level variations along the North Sea coast: Natural variability and anthropogenic change. Cont Shelf Res 19

96. Lauer W (1981) Klimawandel und Menschheitsgeschichte auf dem mexikanischen Hochland. Akademie der Wissenschaften und Literatur Mainz. Abhandlungen der mathematisch–naturwissenschaftlichen Klasse 2

97. Lin JWB, Neelin JD (2002) Considerations for stochastic convective parameterization. J Atmos Sci 59:959-975

98. Lin JWB, Neelin JD (2000) Influence of a stochastic moist convective parameterization on tropical climate variability. Geophys Res Lett 27:3691-3694

99. Lionello P, Nizzero A, Elvini E (2003) A procedure for estimating wind waves and storm-surge climate scenarios in a regional basin: The Adriatic Sea case. Clim Res 23:217-231

100. Livezey RE (1995) The evaluation of forecasts. In: von Storch H, Navarra A (eds) Analysis of climate variability: Applications of statistical technique. Springer, Berlin Heidelberg New York

101. Lohmann U, Roeckner E (1995) Influence of cirrus cloud radiative forcing on climate and climate variability in a general circulation model. J Geophys Res 100D:16305-16323

102. Lorenz EN (1963) Deterministic nonperiodic flow. J Atmos Sci 20:130-141

103. Lorenz EN (1967) The nature and theory of the general circulation of the atmosphere. World Meteorological Organization

104. Luterbacher J (2001) The Late Maunder Minimum/1675–1715 – Climax of the Little Ice Age in Europe. In: Jones PD, Ogilvie AEJ, Davies TD, Briffa K (eds) History and climate: Memories of the future? Kluwer Academic Press, New York Boston London

105. Luterbacher J, Rickli R, Xoplaki E, Tinguely C, Beck C, Pfister C, Wanner H (2001) The Late Maunder Minimum (1675–1715) – a key period for studying decadal scale climatic change in Europe. Clim Change 49:441-462

106. Luterbacher J, Xoplaki E, Dietrich D, Rickli R, Jacobeit J, Beck C, Gyalistras D, Schmutz C, Wanner H (2002) Reconstruction of sea level pressure fields over the Eastern North Atlantic and Europe back to 1500. Clim Dyn 18:545-561 (DOI 10.1007/s00382-001-0196-6)

107. Manabe S, Bryan K (1969) Climate calculations with a combined ocean–atmosphere model. J Atmos Sci 26:786-789

108. Manabe S, Stouffer RJ (1988) Two stable equilibria of a coupled ocean–atmosphere model. J Clim 1:841-866

109. Markham SF (1947) Climate and the energy of nations, 2nd American ed. Oxford University Press, London New York Toronto

110. Marotzke J (1997) Boundary mixing and the dynamics of three-dimensional thermohaline circulations. J Phys Oceanogr 27:1713-1728

111. Marotzke J (1990) Instabilities and multiple equilibria of the thermohaline circulation. PhD thesis. Berichte aus dem Institut für Meereskunde, Kiel 194

112. Marotzke J (1994) Ocean models in climate problems. In: Malanotte-Rizzoli P, Robinson AR (eds) Ocean processes in climate dynamics: Global and mediterranean examples, Kluwer

113. Mesinger F, Arakawa A (1976) Numerical methods used in atmospheric models. Global Atmospheric Research Program (GARP) WMO-ICSU 17

114. Meteorological Service (1995) A first approach towards calculating synoptic forecast charts by Felix M Exner. Historical Note 1, Glasnevin Hill, Dublin 9, Ireland, February 1995

115. Mikolajewicz U, Maier-Reimer E (1990) Internal secular variability in an OGCM. Clim Dyn 4:145-156

116. Monmonier M (1999) Air apparent. The University of Chicago Press, Chicago

117. Morgan (1999) Learning from models. In: Morgan MS, Morrison M (eds) Models as mediators. Perspectives on natural and social science. Cambridge University Press

118. Morrison M (1999) Models as autonomous agents. In: Morgan MS, Morrison M (eds) Models as mediators. Perspectives on natural and social science. Cambridge University Press

119. Munk WH (1950) On the wind-driven ocean circulation. J Meteor 7:79-83

120. Munk W, Wunsch C (1998) Abyssal recipes II: Energetics of tidal and wind mixing. Deep Sea Res 45:1997-2010

121. Murphy A, Brown BG, Chen YS (1989) Diagnostic verification of temperature forecasts. Weather and Forecasting 4:485-501

122. Navarra A (1995) The development of climate research. In: von Storch H, Navarra A (eds) Analysis of climate variability: Applications of statistical techniques. Springer, Berlin Heidelberg New York

123. Nebecker F (1995) Calculating the weather: Meteorology in the twentieth century. San Diego University Press

124. Nitsche G, Wallace JM, Kooperberg C (1994) Is there evidence of multiple equilibria in the planetary-wave amplitude? J Atmos Sci 51:314-322

125. Oberle HJ, von Storch H, Tahvonen O (1996) Numerical computation of optimal reductions of CO_2 emissions in a simplified climate–economy model. Numer Funct Anal Optimiz 17:809-822

126. Olbers DJ, Wenzel M (1989) Determining diffusivities from hydrographic data by inverse methods with applications to the circumpolar current. In: Anderson DLT, Willebrand J (eds) Oceanic circulation models: Combining data and dynamics.

127. Oreskes N, Shrader-Frechette K, Beltz K (1994) Verification, validation, and confirmation of numerical models in earth sciences. Science 263:641-646

128. Pacyna JM, Pacyna EG (2000) Atmospheric emissions of anthropogenic lead in Europe: Improvements, updates, historical data and projections. GKSS Report 2000/3

129. Pedlosky J (1987) Geophysical fluid dynamics. Springer, Berlin Heidelberg New York

130. Peixoto JP, Oort AH (1992) Physics of climate. American Institute of Physics

131. Pekeris CL, Accad Y (1969) Solution of Laplace's equation for the M_2-tide in the world oceans. Phil Trans Roy Soc Lond, Ser A 265:413-436

132. Petersen AC (1999) Philosophy of climate science. Bull Am Met Soc 81:265-271

133. Pfister C (1999) Wetternachhersage. 500 Jahre Klimavariationen und Naturkatastrophen 1496-1995. Haupt, Bern Stuttgart Wien

134. Preisendorfer RW (1988) Principal component analysis in meteorology and oceanography. Elsevier, Amsterdam

135. Pruppacher HR, Klett JD (1978) Microphysics of clouds and Precipitation. Reidel Publishing Co

136. Rahmstorf S (1995) Bifurcations of the Atlantic thermohaline circulation in response to changes in the hydrological cycle. Nature 378:145-149

137. Ray RD (1998) Ocean self-attraction and loading in numerical tidal models. Mar Geod 21:181-192

138. Rebetez M (1996) Public expectation as an element of human perception of climate change. Clim Change 32:495-509

139. Reiter P (2001) Climate change and mosquito-borne disease. Env Health Persp 109, Suppl 1:141-161

140. Richardson LF (1922) Weather Prediction by Numerical Process. Cambridge University Press

141. Robinson AR, Lermusiaux PFJ, Sloan III NQ (1998) Data assimilation. In: Brink KH, Robinson AR (eds) The global coastal ocean. Processes and methods. The Sea, vol 10. John Wiley and Sons, New York

142. Roeckner E, von Storch H (1982) On the efficiency of horizontal diffusion and numerical filtering in an Arakawa-type model. Atmosphere-Ocean 18:239-25

143. Rooth C (1982) Hydrology and ocean circulation. Prog Oceanogr 11:131-149

144. Saloranta TM (2001) Post-normal science and the global climate change issue. Clim Change 50:395-404

145. Schatzmann M (2001) Physical modeling of flow and dispersion. In: von Storch H, Flöser G (eds) Models in environmental research. Springer, Berlin Heidelberg New York

146. Schröter J, Wunsch C (1986) Solution of nonlinear finite difference ocean models by optimization methods with sensitivity and observational strategy analysis. J Phys Oceanogr 16:1855-187

147. Schwartz P (1991) The art of the long view. John Wiley and Sons

148. Schwiderski EW (1980) Global ocean tides. Part I: Global ocean tidal equations. Mar Geod 3:161-217

149. Scott JR, Marotzke J (2002) The location of diapycnal mixing in the meridional overturning circulation. J Phys Oceanogr 32:3578-3595

150. Serres M, Farouki N (eds) (2001) Thesaurus der exakten Wissenschaften. Zweitausendeins, Frankfurt a M

151. Seiler U (1989) An investigation to the tides of the world ocean and their instantaneous angular momentum budgets. Mitt Inst Meereskd Univ Hamb 29

152. Shindell DT, Schmidt GA, Mann M, Rind D, Maple A (2001) Solar forcing of regional climate change during the Maunder Minimum. Science 294:2149-2152

153. Smagorinsky JS (1963) General circulation experiments with the primitive equations I. The basic experiment. Mon Wea Rev 91:99-164

154. Staff members of the Institute of Meteorology, University of Stockholm (1954) Results of forecasting with the barotropic model on an electronic computer (BESK). Tellus 6:139-149

155. Stehr N, von Storch H (eds) (2000) Eduard Brückner – The sources and consequences of climate change and climate variability in historical times. Kluwer Publisher

156. Stehr N, von Storch H (1998) Soziale Naturwissenschaft oder die Zukunft der Wissenschaftskulturen. Vorgänge 37:8-12

157. Stehr N, von Storch H (2000) Von der Macht des Klimas. Ist der Klimadeterminismus nur noch Ideengeschichte oder relevanter Faktor gegenwärtiger Klimapolitik? Gaia 9:187-195

158. Stendel M, Schmith T, Roeckner E, Cubasch U (2002) The climate of the 21st century: Transient simulations with a coupled atmosphere–ocean general circulation model. Danish Climate Centre Report 02-1

159. Stocker TF, Wright DG (1991) Rapid transitions of the ocean's deep circulation induced by changes in surface water fluxes. Nature 351:729-732

160. Stommel H (1948) The westward intensification of wind-driven ocean currents. Trans Am Geophys Union 29:202-206

161. Stommel H (1961) Thermohaline convection with two stable regimes of flow. Tellus 13:224-230

162. Suarez MJ, Schopf PS (1988) A delayed action oscillator for ENSO. J Atmos Sci 45:3283-3287

163. Sündermann J, Vollmers H (1972) Tidewellen in Ästuarien. Wasserwirtschaft 62:1-9

164. Sverdrup HU (1927) Dynamics of tides on the North Siberian shelf, Geofys Publ 4,5

165. Tafferner A, Egger J (1990) Test of theories of lee cyclogenesis: ALPEX cases. J Atmos Sci 47:2418-2428

166. Tahvonen O, von Storch H, von Storch J (1994) Economic efficiency of CO_2 reduction programs. Clim Res 4:127-141

167. Ulbrich U, Ponater M (1992) Energy cycle diagnosis of two versions of a low resolution GCM. Met Atmos Phys 50:197-210

168. Ulbrich U, Speth P (1992) The global energy cycle of stationary and transient atmospheric waves: Results from ECMWF analyses. Met Atmos Phys 45:125-138

169. Ungar S (1992) The rise and relative fall of global warming as a social problem. Sociolog Quarter 33:483-501

170. van Andel T (1994) New views on an old planet. A history of global change. Cambridge University Press, 2nd ed

171. Vemuri V (1978) Modeling of complex systems. Academic Press, New York

172. von Hann J (1903) Handbook of climatology. Part I: General climatology. Macmillan, New York

173. von Storch H (1997) Conditional statistical models: A discourse about the local scale in climate modelling. In: Müller P, Henderson D (eds) Monte Carlo simulations in oceanography. Proceedings Áha Hulikoá Hawaiian Winter Workshop, University of Hawaii at Manoa, January 14-17

174. von Storch H (1999) The global and regional climate system. In: von Storch H, Flöser G (eds) Anthropogenic climate change. Springer, Berlin Heidelberg New York

175. von Storch H, Costa-Cabral M, Hagner C, Feser F, Pacyna J, Pacyna E, Kolb S (2003) Four decades of gasoline lead emissions and control policies in Europe: A retrospective assessment. The Science of the Total Environment (STOTEN) 311:151-176

176. von Storch H, Frankignoul C (1998) Empirical modal decomposition in coastal oceanography. In: Brink KH, Robinson AR (eds) The global coastal ocean. Processes and methods. The Sea, vol 10. John Wiley and Sons, New York

177. von Storch H, Güss S, Heimann M (1999) Das Klimasystem und seine Modellierung. Eine Einführung. Springer, Berlin Heidelberg New York

178. von Storch H, Hasselmann K (1996) Climate variability and change. In: Hempel G (ed) The ocean and the poles. Grand challenges for European cooperation. Fischer, Jena Stuttgart New York

179. von Storch H, Langenberg H, Feser F (2000) A spectral nudging technique for dynamical downscaling purposes. Mon Wea Rev 128:3664-3673

180. von Storch H, Stehr N (2000) Climate change in perspective. Our concerns about global warming have an age-old resonance. Nature 405:615

181. von Storch H, von Storch JS, Müller P (2001) Noise in the climate system – Ubiquitous, constitutive and concealing. In: Engquist B, Schmid W (eds)

Mathematics unlimited – 2001 and beyond, part II. Springer, Berlin Heidelberg New York

182. von Storch H, Zwiers FW (1999) Statistical analysis in climate research. Cambridge University Press

183. von Storch JS (1999) Natural climate variability and concepts of natural climate variability. In: von Storch H, Flöser G (eds) Anthropogenic climate change. Springer, Berlin Heidelberg New York

184. von Storch JS (2000) Signatures of air–sea interactions in a coupled atmosphere–ocean GCM. J Clim 13:3361-3379

185. Walker GT, Bliss EW (1932) World weather V. Mem R Met Soc 4:53-84

186. Washington WM, Parkinson CL (1986) An introduction to three-dimensional climate modelling. University Science Books

187. Wiggins S (1990) Introduction to applied nonlinear dynamical systems and chaos. Springer, Berlin Heidelberg New York

188. Wilks DS (1995) Statistical methods in atmospheric sciences. Academic Press, San Diego

189. Wright PB (1984) On the relationship between indices of the southern oscillation. Mon Wea Rev 1112:1913-1919

190. Wright PB (1985) The southern oscillation – an ocean–atmosphere feedback system. Bull Am Met Soc 66:398-412

191. Wunsch C (1996) The ocean circulation inverse problem. Cambridge University Press, Cambridge

192. Wunsch C (1998) The work done by the wind on the oceanic general circulation. J Phys Oceanogr 28:2332-2340

193. Zahel W (1978) The influence of solid earth deformations on semidiurnal and diurnal oceanic tides. In: Brosche P, Sündermann J (eds) Tidal friction and the earth's rotation. Springer, Berlin Heidelberg New York

194. Zinke J, von Storch H, Müller B, Zorita E, Rein B, Mieding HB, Miller H, Lücke A, Schleser GH, Schwab MJ, Negendank JFW, Kienel U, González-Ruoco JF, Dullo C, Eisenhauser A (2004) Evidence for the climate during the Late Maunder Minimum from proxy data available within KIHZ. In: Fischer H, Kumke T, Lohmann G, Flöser G, Miller H, von Storch H, Negendank JFW (eds) The KIHZ project: towards a synthesis of Holocene proxy data and climate models. Springer, Berlin Heidelberg New York (in press)

195. Zorita E, González-Rouco F (2003) Are temperature-sensitive proxies adequate for North Atlantic Oscillation reconstructions? Geophys Res Let 29:481-484

196. Zorita E, von Storch H, González-Rouco F, Cubasch U, Luterbacher J, Legutke S, Fischer-Bruns I, Schlese U (2003) Simulation of the climate of the last five centuries. GKSS Report 2003/12

197. Zwiers FW (1999) The detection of climate change. In: von Storch H, Flöser G (eds) Anthropogenic climate change. Springer, Berlin Heidelberg New York

Index

Printing: Mercedes-Druck, Berlin
Binding: Stein+Lehmann, Berlin